Advances in Industrial Control

Other titles published in this series:

Pietro J. Dolcini · Carlos Canudas de Wit
Hubert Béchart

Dry Clutch Control for Automotive Applications

Pietro J. Dolcini, Dr.
Centre Technique Renault de Lardy
1 Allée Cornuel
91750 Lardy
France
pietro.dolcini@renault.com

Hubert Béchart, Prof.
RENAULT
1 Avenue du Golf
78288 Guyancourt
France
hubert.bechart@renault.com

Carlos Canudas de Wit
Director of Research at the CNRS
Département d'Automatique de Grenoble
GIPSA-Lab, UMR CNRS 5216, BP. 46
38402 Saint Martin d'Hères
France
carlos.canudas-de-wit@inpg.fr

ISSN 1430-9491
ISBN 978-1-4471-2560-0   ISBN 978-1-84996-068-7(eBook)
DOI 10.1007/978-1-84996-068-7
Springer London Dordrecht Heidelberg New York

British Library Cataloguing in Publication Data
A catalogue record for this book is available from the British Library

Advances in Industrial Control

Professor (Emeritus) O.P. Malik
Department of Electrical and Computer Engineering
University of Calgary
2500, University Drive, NW
Calgary, Alberta
T2N 1N4
Canada

Professor K.-F. Man
Electronic Engineering Department
City University of Hong Kong
Tat Chee Avenue
Kowloon
Hong Kong

Professor G. Olsson
Department of Industrial Electrical Engineering and Automation
Lund Institute of Technology
Box 118
S-221 00 Lund
Sweden

Professor A. Ray
Department of Mechanical Engineering
Pennsylvania State University
0329 Reber Building
University Park
PA 16802
USA

Professor D.E. Seborg
Chemical Engineering
3335 Engineering II
University of California Santa Barbara
Santa Barbara
CA 93106
USA

Doctor K.K. Tan
Department of Electrical and Computer Engineering
National University of Singapore
4 Engineering Drive 3
Singapore 117576

Professor I. Yamamoto
Department of Mechanical Systems and Environmental Engineering
The University of Kitakyushu
Faculty of Environmental Engineering
1-1, Hibikino,Wakamatsu-ku, Kitakyushu, Fukuoka, 808-0135
Japan

Series Editors' Foreword

The series *Advances in Industrial Control* aims to report and encourage technology transfer in control engineering. The rapid development of control technology has an impact on all areas of the control discipline. New theory, new controllers, actuators, sensors, new industrial processes, computer methods, new applications, new philosophies , new challenges. Much of this development work resides in industrial reports, feasibility study papers and the reports of advanced collaborative projects. The series offers an opportunity for researchers to present an extended exposition of such new work in all aspects of industrial control for wider and rapid dissemination.

The *Advances in Industrial Control* series began in 1992, and since then has published only one volume from the field of vehicle control. Even that seminal work by P. Kachroo and K. Ozbay entitled *Feedback Control Theory for Dynamic Traffic Assignment* (ISBN 978-1-85233-059-0, 1998) was on traffic management rather than vehicle control *per se*. This seems quite an important omission from the series when the recent growth and influence of control-system techniques in many diverse aspects of automotive vehicle control is considered. A recent look at international control conferences identified a wide range of topics in this field including sessions and papers on:

- control of braking systems;
- engine control and engine-health management;
- ignition-system control and novel developments;
- control of gasoline and Diesel-fueled engines;
- control of electric automotive motors;
- electric-vehicle systems;
- hybrid-vehicle systems;
- *in-traffic* control systems;
- control systems for autonomous land vehicles;

- multi-vehicle control - simple configurations, *e.g.*, spacing;
- multi-vehicle control - complex configurations, *e.g.*, co-operative maneuvering.

Clearly, there is much control systems research activity in the field, and the Series Editors are pleased to introduce this first *Advances in Industrial Control* series monograph on automotive vehicle control. This sharply focused volume entitled Dry Clutch Control for Automotive Applications by P.J. Dolcini, C. Canudas de Wit and H. Béchart, presents some new ideas for enhancing powertrain driving comfort during a standing-start or in gear-shifting maneuvers using the clutch element of the drivetrain. The monograph opens by describing the range of practical and aesthetic constraints that limit control engineering design freedom.

Powertrain packaging constraints, such as the necessary preservation of interior car capacity, and the spatial constraints emerging from front collision robustness requirements, are given. Other practical constraints are:

- minimum required ground clearance;
- wheel steering movement volume limitations;
- pedestrian collision test requirements;
- style requirements for the vehicle front design.

Within the context of these constraints, the authors pose this key question: what can be done with clutch design and control to enhance powertrain drive comfort? Some answers are presented in the six succinct chapters of the monograph. These are arranged in two parts. Part I covers the mechanical description and mathematical modeling of the drivetrain and Part II pursues several aspects of the control solution for clutch design. Included are a chapter on a synchronized clutch- assist system and a chapter presenting experimental results from a Renault Clio prototype vehicle. A chapter on 'Conclusions and Open Questions' ends the book.

The monograph will be of obvious interest to both automotive engineers and control engineers, and also to researchers and academics with an interest in automotive vehicle control problems. Those interested in demanding real-world control applications may also find this monograph provides a suitably challenging set of problems for new design techniques.

Fortunately, the depth of modeling presented will enable a wide range of readers, be they researchers, academic or students, to try their skills on the clutch control problems described. As was stated at the beginning of this Foreword, there is significant international research activity evident at the interfaces between many aspects of automotive engineering, and control system design. Consequently, because the series editors are always seeking new monographs to create a corpus of contributions to important and active fields and in anticipation of two new volumes on automotive vehicle control expected to appear in the series soon, it is pleasing to initiate the *Advances in Industrial Control* series contribution to this field with this excellent text from P. J. Dolcini, C. Canudas de Wit and H. Béchart.

Industrial Control Centre   *M.J. Grimble*
Glasgow   *M.A. Johnson*
Scotland, UK
2009

Contents

Part II Dry Clutch Engagement Control

1

Introduction

The driving comfort perceived by the driver is as important for the commercial success of a vehicle as its dynamic performances or its fuel efficiency. Unfortunately, though, this element is much more difficult to measure and to master than the last two. During the conception and the final fine tuning of a new car substantial efforts are made to insure a correct level of the different comfort performances. These performances depend on most of the components of the vehicle and impose strict constraints on the technical choices available to the engineer.

Developments in the car's overall architecture design tend toward a reduction in the powertrain volume. The first issue underlying this trend involves habitability requirements. The concept of interior spaciousness, first introduced by the innovative monospace vehicle *Renault Espace* in the early 1980s and now slowly migrating to conventional vehicles, aims at maximizing the space dedicated to passengers without increasing vehicle length or width. This can be achieved by raising the roof, flattening the floor, reducing the amount of room occupied by the rear axle and, of course, reducing the space dedicated to the powertrain.

The second constraint is due to front-collision robustness requirements. In order to pass crash tests successfully, particularly the so-called *Thatcham* or *Danner test*[1] used by insurance companies to calculate repair-cost ratios, modern vehicles must provide a dead volume between the front bumper beam and the engine (generally delimited by the cooling radiator). This volume is designed to tolerate bumper-beam deformation induced by the shock and its value is generally related to crash-test performance.

Other constraints on powertrain packaging are the minimum ground clearance requirements, the volume dedicated to wheel steering movements (which is

---

[1] This European 15 km/h front and back crash protocol, mainly designed to test the bumper's effectiveness, is much more stringent than the equivalent American 8 km/h IIHS test.

tending to increase due to the current trend for rim diameters to increase), pedestrian collision tests and style requirements concerning the vehicle front design.

The clutch is a key element for the powertrain driving comfort during standing-start and gear-shifting maneuvers in manual transmission cars, which constitute the vast majority of the European automotive park. Furthermore, driven by a strong pressure for fuel efficiency, automated manual transmission and dual-clutch transmission systems are introduced in historically automatic transmission markets such as North and South Americas and far eastern countries like Korea and Japan. These systems, essentially improved robot-controlled manual transmission gearboxes, make use of one or more clutches during standing-start and gear-shifting maneuvers and, therefore, share most driving comfort issues with their manual counterparts.

Now let us look at how the design of a manual transmission dry clutch assembly, composed by the clutch itself and its actuation system, are effected by the previously highlighted powertrain constraints.

The maximum torque output of passenger car Diesel engines has been consistently rising over the past few years. This is due to a constantly increasing demand for performance, driven by the constant increase of vehicle mass, and the higher specific torque obtained with new Diesel technologies such as high-pressure direct injection and high boost pressure.

Since the clutch, when fully closed, must ensure slip-free transmission of the engine torque, its capacity requirements must be increased. There are two methods that comply with this requirement: increasing the diameter of the friction discs or increasing the normal force exerted by the washer spring on the discs.[2] Because of the powertrain packaging constraints discussed above, increasing the diameter of the friction discs is no longer an option because there is no room available for a larger clutch. Besides the size constraint, a greater diameter increases the discs' moment of inertia inducing shocks when shifting gear and downgrading the driver's comfort.

The alternative solution, *i.e.* to increase the normal force exerted on the discs by modifying the washer spring's stiffness, has two main drawbacks. The first is that it increases the overall effort to be provided by the driver on the clutch pedal. A limit has been reached insofar as concerns clutch pedal effort so that any increase means a loss of driver comfort due to the difficulty in producing the effort and the strain induced by repeated clutch operation (notably in traffic jam driving). The second is that it reduces the progressiveness of clutch command (by reducing the slip range of the pedal stroke) making satisfactory

---

[2] The maximal torque a clutch can transmit is: $\Gamma_c = 2n\mu R_c F_{max}$ where n is the number of clutch disks, μ the garniture friction coefficient, R_c the mean friction radius and F_{max} the normal force exerted by the washer plate on the friction surfaces in rest position. $\mu \cong 0.4$ is the technologically feasible value. $n > 1$ is an expensive solution usually limited to sports cars.

clutch engagement harder to achieve for a normal driver (more attention is required to avoid torque shocks or even engine stall).

The only solution that avoids these drawbacks for a normal clutch system would be to lengthen the pedal stroke. Unfortunately, this solution is inadmissible because of ergonomic constraints and also because it would consume precious centimeters in the driver's seat longitudinal position, reducing the interior volume.

This situation motivates the interest of both automotive constructors and their suppliers in breakthrough solutions including actively aided clutch activation systems, clutch-by-wire and automated manual transmission or dualclutch transmission systems. This book, based on the research work done in collaboration between the automotive constructor Renault S.A. and the GIPSA-Lab in Grenoble, gives a control-oriented analysis of the dry clutch engagement problem that is at the heart of all the previously listed breakthrough solutions.

The following material is divided into two parts: the first gives a description of the mechanical elements of the driveline, their mathematical modeling and their interaction during a standing-start and gear-shifting maneuvers before analyzing the clutch-related driving comfort. The second part of the book discusses the dry clutch engagement control problem in the case of a synchronization-assistance scheme in which the driver directly controls the clutch position but for the very last part of the engagement and in the case of a fully automatic engagement scheme. Finally, experimental results based on a Clio II 1.5dCi prototype equipped with a five-speed Renault AMT JH gearbox are presented.

Part I

Mechanical System and Comfort Requirements

2

Powertrain

2.1 Brief Mechanical Description

2.1.1 Elements of the Engine Block

The engine block and the driveline, together forming the powertrain, are the mechanical elements assuring the vehicle's main function, *i.e.* to move. Several architectures are available for the powertrain; in this work we will only consider the manual transmission (MT) and automated manual transmission (AMT) systems. Starting from the engine and moving toward the wheels the elements of the powertrain are:

- engine;
- flywheel or dual-mass flywheel (DMFW);
- dry clutch;
- gearbox and differential;
- transmission shafts; and
- tires.

In the following sections of this chapter a brief review of the different elements of the powertrain with some details about their structure and how they work will be given in order to allow a better understanding of the challenges involved in the clutch control. Finally, two models, a detailed simulation model and a simpler control model, will be presented. For more detailed information, particularly about the engine control, the reader is invited to consult a reference book on the subject like [21].

2.1.2 Engine

Gasoline and Diesel Engines

In both gasoline and Diesel engines power is generated through a four-stroke cycle performed in two complete revolutions. The two engines share the four-stroke division of the cycle, namely: intake, compression, power and exhaust strokes, but differ in the way the air fuel mixture is ignited. In the gasoline engine the ignition is triggered through a spark while in the Diesel engine the mixture simply auto-ignites due to the temperature and pressure conditions in the combustion chamber. The means of creating the air fuel mixture introduce an important technical difference. Usually, gasoline engines sport an indirect injection meaning that the gasoline is injected in the manifold before the admission valve; therefore, during the admission stroke, the cylinder is filled with an air fuel mixture. Diesel engines, instead, usually have a direct injection, *i.e.* the fuel is directly injected in the combustion chamber during the compression phase.

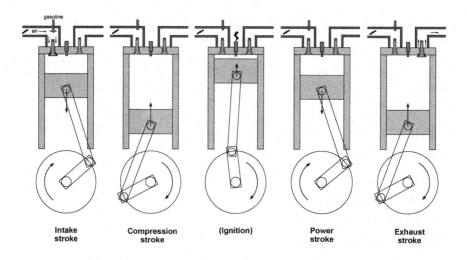

Figure 2.1. Gasoline engine four-stroke cycle

During the first stroke of an indirect injection gasoline engine the piston descends from the top dead center (TDC) to the bottom dead center (BDC). The admission valve is open and the cylinder is filled with an air fuel mixture coming from the manifold where the gasoline has been injected just before opening the valve. The air mass allowed to flow in the cylinder is controlled by the throttle plate, a butterfly valve partially choking the intake airflow and, thus, lowering the pressure in the manifold. The amount of gasoline injected is a function of the air mass in order to assure a 14.7 : 1 fuel to oxygen stoichiometric ratio, *i.e.* a perfectly balanced combustion neither too lean or too rich in fuel. The factory pre-set values obtained through engine calibration are corrected online by feedback on the λ sensor readings measuring the oxygen partial pressure in the exhaust gasses. During the compression stroke the intake valve is closed and the piston, following its movement from the BDC to the TDC, compresses the mixture. A few degrees before the TDC the combustion is triggered by a spark delivered by a plug. The angular position of the crankshaft relative to the TDC at which the spark is triggered, usually ranging between -40 and 10 degrees, is called *spark advance* and allows for control of the torque delivered by the engine during the power stroke during which the piston moves from TDC to the BDC. For evident reasons of fuel efficiency the spark advance is usually set around the optimal angle of about -25 degrees, delivering the maximum torque output for a given quantity of gasoline. The last stroke, the exhaust stroke, allows for the evacuation of the spent gasses through the exhaust valve while the piston returns to the TDC ready for a new cycle.

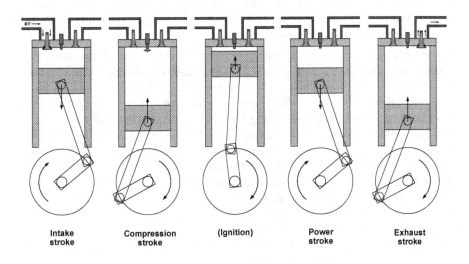

| Intake stroke | Compression stroke | (Ignition) | Power stroke | Exhaust stroke |

Figure 2.2. Diesel engine four-stroke cycle

Compared to the cycle of a indirect injection gasoline engine the cycle of a Diesel engine shows some differences in the first two strokes. During the admission stroke the cylinder is filled only with fresh air without any control on the intake flow since the throttle plate is absent. During the compression stoke the fuel is injected in the cylinder; the shape of the intake pipes, of the piston's head and the angle of injection are designed to create a bubble of stoichiometric mixture at the center of the combustion chamber thus allowing a globally lean mixture, due to the fresh air surrounding the bubble, while having a locally stoichiometric mixture where the combustion happens. This arrangement both reduces noxious emission and increases fuel efficiency. At the end of the compression stroke vaporization, pressure and temperature conditions for auto-ignition are met thus allowing for a power stroke. The rest of the cycle proceeds exactly like the previous one.

Fast, Slow and Negative Torque

The torque output during the power stroke is a function both of the quantity of injected fuel and the ignition point. The following discussion is strictly valid only for older atmospheric engines without any gas recirculation, multiple injection or camshaft dephaser devices. The new-generation engines control is by far more complex and gives better performances, but for most clutch-related comfort purposes we can limit out attention to the older, more simple case.

The gasoline engine has three control actuators: the throttle plate, the injectors and the spark plugs. The need for a air fuel stoichiometric ratio imposes a constraint on the fuel injection, thus effectively reducing the control inputs to the throttle plate and the spark plug.

The throttle plate controls the intake pressure in the manifold and thus, indirectly, the amount of fuel injected. This pressure sets an upper limit to the torque output that is reached if the mixture is ignited with an optimal lead angle. A delayed ignition, called *advance reduction*, allows a reduction the effective torque output by degrading the conversion efficiency of the engine. This controlled reduction of the engine efficiency can be explained by the slow dynamic of the intake pressure. In order to allow a better response of the engine a small *torque reserve* is made, meaning that an intake pressure slightly higher that what is needed is used in combination with a less-than-optimal spark advance for compensation. When faced with a request of a sudden increase of the output torque the control engine can increase the lead angle to the optimum while waiting for the intake pressure to rise thanks to the opening of the throttle plate. On the other hand when faced with a request for a sudden decrease of the output torque the engine control chokes the inflow with the throttle plate and reduces the lead angle while waiting for the intake pressure to drop. In engine control lingo *slow torque* refers to the potential torque that the intake pressure could generate if the air fuel mixture is ignited at the optimal lead angle, *fast torque*, instead, refers to the actual output torque that could be lower than the previous value due to a non optimal lead angle. The slow torque has a characteristic time of about 0.04 s while the fast torque can take any value between zero and the slow torque every TDC.[1]

---

[1] A strong reduction of the lead angle causes the combustion to complete in the exhaust pipes. Since the exhaust system is not designed to withstand such high

The Diesel engine, on the other hand, has as control only the quantity of injected fuel and its timing since it lacks both the throttle plate[2] and spark plug. Thanks to the clever arrangement of the intake pipes, combustion chamber shape and injector angles leading to a globally lean, locally stoichiometric air fuel mixture almost no constraint is put on the quantity of injected fuel. The output torque control is therefore much simpler being reduced to the fast torque signal.[3]

The internal combustion engine can be thought of as a pump taking air from the intake circuit and forcing it in the exhaust pipes. If no torque is generated during the power stroke, this pumping work, together with the internal friction losses, creates a net negative torque of about −50 Nm.

For the rest of this presentation we will denote Γ_e the mean net engine output torque over one half revolution (from TDC to TDC) ranging from a maximum of about 200 Nm, depending on the engine characteristics, to a minimum of about −50 Nm.

Throttle Look-up Table

The static relation giving the engine torque target for the engine control unit as a function of the throttle pedal position and the engine speed is called the *throttle look-up table*. This target value can be further modified by the engine control unit strategies aiming at, for example, reducing obnoxious emissions, increasing comfort or avoiding engine stalling.

The iso-power contours in the torque-engine speed plan are the starting point for filling in this table; these initial values are then modified to take into account ergonomic and performance requirements and, finally, fine tuned directly on the vehicle.

2.1.3 Flywheel and Dual-mass Flywheel

Flywheel

Of the engine's cycle four strokes only the power stroke delivers a positive torque, the other three having a negative balance due to friction and compression and pumping work. The phase shift between the different pistons

---

temperatures heavy reductions of the lead angle are possible only for a limited time.

[2] Actually some diesel engines have something similar to a throttle plate on the intake conduct but it is only used to choke down the engine rapidly when the key contact is broken.

[3] For software-compatibility reasons the distinction between fast and slow torque is artificially kept even in Diesel engine. Although in this chapter the subject won't be further developed, intake pressure of turbocharged engines can be controlled by means of a turbocharger cut-off valve called a waste-gate.

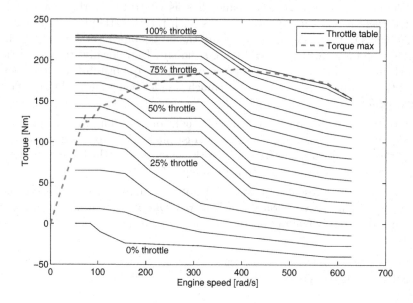

Figure 2.3. Throttle look-up table for a gasoline engine traced in solid black together with the maximum output torque for a given engine speed traced in dashed gray

assures a rough balance of the output torque. Considering the most common case in European cars of a four-cylinder four-stroke engine, in fact, one piston is always completing a power stroke while another is finishing its compression stroke and getting ready for a new power stroke (Figure 2.4).

The instantaneous output torque resulting from the concurrent action of the four pistons shows peaks, betraying the controlled explosion of an internal combustion engine. These peaks induce oscillations of the engine speed called engine acyclicity. In order to limit these oscillations a flywheel, *i.e.* a solid cast iron wheel having a big rotational inertia, is added to one end of the crankshaft.

Besides reducing the engine-speed oscillations the flywheel also performs three auxiliary functions:

- It serves as a reduction gear for the cranking-up of the engine.

- It has on its outer perimeter a toothed target used for calculating both the engine revolution speed and the crankshaft angle for ignition and injection timing.

- The gearbox-facing side is used as a friction surface for the clutch disk.

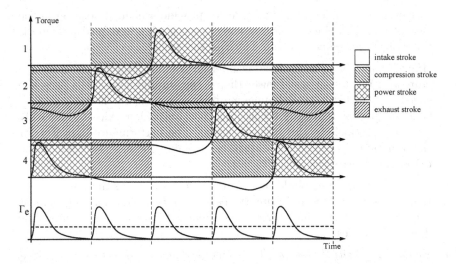

Figure 2.4. Schematic representation of the output torque for each piston in a four cylinder engine having a standard 1342 ignition pattern. The last line shows the total instantaneous output torque and its Γ_e mean value over an half-revolution.

Dual-mass Flywheel

The dual-mass flywheel (DMFW) is a flywheel composed of two disks connected by a damper spring device. This evolution of the classic flywheel is designed to filter out the engine acyclicity before the driveline. In the case of a simple flywheel this filtering is performed by the damper spring system in the clutch disk.

2.1.4 Dry Clutch

Clutch System

The *clutch system* is the set of mechanical elements allowing to smoothly make and break the connection between the engine and the driveline. This system is composed of a connecting element and its control system. Several technical solutions are available for these two elements, in this chapter only the dry single-disk clutches with an hydraulic actuator will be considered. This configuration, by far the most common for MT and AMT cars, is the one used on all vehicles produced by Renault.

The clutch assures four main functions:

- *Decoupling of the Engine and the Driveline* This decoupling can be either of short duration, like, for example, while performing a standing-start or a gear shift, or much longer in order to provide a neutral position in the case of a AMT vehicle. The residual torque in the decoupled position is the main performance indicator for this function.

- *Allowing standing-starts* Since an internal combustion engine cannot operate at a revolution speed lower than a certain minimum, called the idle speed, at which the available engine torque is equal to the internal friction and pumping losses,[4] a clutching mechanism is needed to smoothly launch the vehicle to this minimal speed. The performance indicator for this function is the clutch's *dosability*, meaning the ease with which the driver can control the clutch torque.

- *Easing Gear Shifting* While gear-shifting the clutch eases the synchronization of the crankshaft and primary gearbox shaft speeds. The engagement is quite short but the high torque levels reached can lead to uncomfortable driveline oscillations.

- *Engine Acyclicity Filtering* The engine acyclicity causes torsional vibrations of the crankshaft that, if not filtered out, are transmitted through the driveline to the vehicle body. In order to prevent this a system of damping springs is mounted on the clutch disk. Due to the increasing need of acyclicity filtering, the more powerful engines are equipped with a DMFW that assures a better filtering action. In this latter case the clutch disk presents no damping springs.

Hydraulic Actuator

The hydraulic actuator in an MT vehicle connects the clutch pedal to the clutch washer-spring fingers through a hydraulic circuit composed of a master cylinder called a concentric master cylinder (CMC) directly connected to the pedal, several pipe sections one of which is flexible in order to allow for the movement of the engine on its suspensions, an optional vibration filter and, finally, a slave cylinder called concentric slave cylinder (CSC) that, placed between the gearbox carter and the clutch, pushes directly on the diaphragm fingers.

This hydraulic circuit also assures an effort reduction through the combined effect of the surface ratio between the CMC and CSC and the lever effect given by the clutch pedal. An additional compensation spring is also present in order to further reduce the force necessary for opening the clutch.

In an AMT vehicle the CSC position is directly controlled by the gearbox control unit through an electro-valve. Since no effort is required by the driver neither an effort-reduction nor a compensating spring are used.

---

[4] Actual idle speed is set slightly higher than this limit for robustness reasons.

Clutch

Modern clutches are the result of a long technical evolution begun with sliding transmission belts used for connecting steam textile mills at the start of the industrial revolution. The clutch was first introduced in the automotive industry, together with other paramount technologies like the battery ignition system, the spark plug, the carburetor, gearshift and water cooling, by Karl Benz with his patented Motorwagon in 1885. Since the 1960s dry clutch designs are based on a single friction disk compressed by a washer spring.[5] The single-disk design offers a great compactness, very important for transversal engine architectures where the engine, the clutch, the gearbox and the differential have to fit in between the wheels.

The flywheel is fixed on the crankshaft that revolves at the engine speed. The clutch external structure, the washer spring and the pressure plate are screwed to the flywheel. The clutch torque is generated by the friction of the friction material pads on each side of the clutch disk against the flywheel and the pressure plate. The clutch disk is fixed at the end of the gearbox primary shaft and transmits the generated torque to the driveline. The disk itself presents spring dampers for filtering the engine acyclicity and a flat spring between the friction material pads. The non-linear stiffness of this flat spring has a paramount role in the *dosability* performances[6] of the clutch.

At rest the washer spring crushes the friction pads and the flat spring between the flywheel and the pressure plate with a force F_0 of about 400 N for a 200 mm clutch designed for a 160 Nm peak torque engine. This force, called *clutch pre-charge*, sets the maximal torque the clutch can deliver, which is proportional to the clutch disk diameter and the applied pressure. The CSC piston exerts an axial force on the prongs on the internal diameter of the washer spring reducing the force on the pressure plate until a complete liberation of the clutch disk occurs.

The non-linear stiffness characteristic of the washer spring has a dip in the middle; a clever choice of the shape of this characteristic, set by the dimensions of the washer spring, matched by a corresponding flat spring allows to strongly reduce the force needed to fully open the clutch.

---

[5] The washer spring is basically a truncated metal cone used as an axial spring. Along the internal diameter several wide cuts are made, the resulting prongs called clutch fingers, are used as a leverage for loosening the spring and thus control the opening of the clutch.

[6] The *dosability* of a clutch is a manual transmission comfort parameter linked to the ease with which the driver can control the transmitted torque through the clutch pedal.

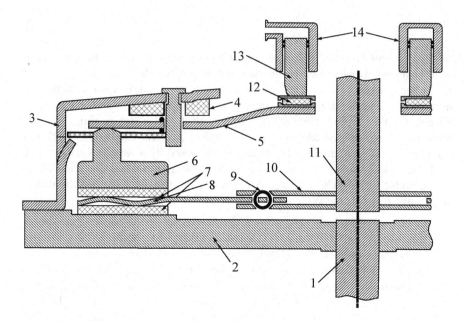

Figure 2.5. Clutch structure, axial cut. 1 crankshaft 2 flywheel 3 clutch external structure 4 wear-compensation system 5 washer spring 6 pressure plate 7 friction pads 8 flat spring 9 spring damper 10 clutch disk 11 gearbox primary shaft 12 needle roller bearing 13 concentric slave cylinder (CSC) piston 14 concentric slave cylinder (CSC)

How the Clutch Works

The discussion in this section is based on the mechanical analysis of the clutch made in the VALEO technical documentation [23].

The starting point in this analysis is Figure 2.6.

At rest (clutch completely engaged) no force is exerted by the CSC on the washer spring fingers ($f = 0$). The washer spring is squashed between the pressure plate and the clutch external structure. This constraint force, called the *pre-charge force*, sets the maximal torque the clutch can deliver.

The axial compression force F_n of a washer spring is determined by its constraint-free shape, its axial compression and the characteristics of the metal composing the spring itself. This force can be estimated by the formula of Almen and László

$$F_n = \frac{4EC}{1 - \nu^2} \frac{e\delta}{D^2} \left((h - \delta) \left(h - \frac{\delta}{2} + e^2 \right) \right),$$

with

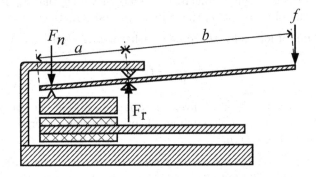

Figure 2.6. Forces acting on the washer spring

$$C = \pi \left(\frac{D}{D-d} \right)^2 \left(\frac{D+d}{D-d} - \frac{2}{\ln(D/d)} \right),$$

where D is the external diameter of the washer spring (Figure 2.7), d the internal one, E the elasticity module of the metal, e the thickness of the washer spring, h height if the truncated cone defined by the unconstrained washer spring, δ the axial deformation of the cone with respect to its unconstrained height, and ν the Poisson coefficient.

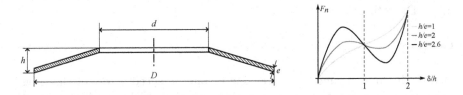

Figure 2.7. Geometry of a washer spring and its stiffness curve for several values of h/e ratio

If the h/e ratio is greater than 1.5 the stiffness characteristic $F_n(\delta/h)$ has a negative derivative in a neighborhood of the $\delta/h = 1$ inflection point. This peculiarity is used for reducing the force needed to operate the clutch. If we assume a perfect rigidity of all the clutch elements except for the washer spring the equilibrium of moments along the radial direction gives

$$F_n = F_0 - \frac{b}{a}f \quad \text{or} \quad f = (F_0 - F_n)\frac{a}{b},$$

where F_0 is the pre-charge force due to the squashing of the washer spring, F_n the normal force exerted on the pressure plate, f the force CSC piston applies to the washer spring's fingers, a and b the leverage that F_n and f have respective to the pivot point on the clutch external structure.

The equilibrium of forces for the whole washer-spring gives

$$F_r = \frac{a}{b+a}f - F_0.$$

F_r changes sign for $f \geq \frac{b+a}{a}F_0$, physically this implies that the washer-spring is turned inside out and completely frees the pressure plate assuring a complete disengagement of the clutch.

Under the simplifying hypothesis of perfect rigidity, the washer-spring fingers move only after the normal force F_n is brought to zero and the clutch disk is completely free. Since it's quite ergonomically difficult to control a movement-free force, a flat spring is introduced between the two friction pads. When a force f is applied to the washer-spring fingers the pressure plate and the washer-spring find a new equilibrium position between F_n and the force exerted by the flat spring.

The stiffness characteristics of the washer and flat springs are carefully chosen in order to have an almost constant f over the greatest possible range of movement of the washer-spring's fingers. The final result is a normal force on the friction surfaces, and thus a transmitted torque, that is essentially a function of just the x_b movement of the washer-spring's fingers

2.1.5 Driveline

Gearbox

Due to the limited range of the engine revolution speed a way of changing the reduction ratio between the crankshaft and the wheels is needed. Several devices have been introduced to assure this function. Seen from the driver's side three main interfaces are available: a completely manual gearshift, a driver triggered automated gearshift and, finally, a completely automated gearshift.[7] Mechanical engineers, instead, classify gearboxes following their working principle:

- *Manual Transmission (MT)* The standard transmission type for European cars. Several discrete reduction ratios are obtained through the selection of coupled gears. During a gearshift the driveline is disconnected from the engine by opening the clutch, leading to a torque interruption.

---

[7] In order to allow for engine braking while descending long steep roads a gear-selection mechanism is present even in the case of completely automated gearshift.

- *Automated Manual Transmission (AMT)* This is a niche solution, more common on sport cars. An hydraulic or, less frequently, an electric actuator is coupled with a standard MT transmission. Both the gear selection and the clutch are controlled by the actuator; gearshift can be either completely automatic or driver triggered. Since an MT transmission is used, gearshift induces a torque interruption.

- *Direct Shift Gearbox (DSG) or Dual-clutch Transmission (DCT)* This is a quite rare solution due to its complexity, actually licensed to the Volkswagen group. DSG is an improvement over AMT aiming to avoid the torque interruption and speed up the gearshift operation. Even and odd gears are placed on two separate shafts each having its own clutch. During a gearshift the clutch on the old gear's shaft is opened while simultaneously closing the clutch on the new gear's shaft, thus allowing for a smooth ratio change.

- *Automatic Transmission (AT)* This is the standard transmission type outside Europe. Discrete reduction ratios are assured by epicyclical trains whose shafts are controlled by small on-off clutches. The sudden speed changes induced by this arrangement are smoothed out by a torque converter, basically composed of two facing turbines dipped in oil. Since the driveline is always connected no torque interruption is present even if torque jumps can be induced by a poor control of the converter slipping speed.

- *Continually Variable Transmission (CVT)* Reduction is assured by a belt running on two opposite cones. The belt sliding along the cones' axes gives a gradually changing reduction ratio. No torque interruption is present.

This research concerns the clutch-related comfort and therefore will concentrate only on the first two solutions even if the standing-start analysis is also valid for DSG/DCT gearboxes.

Differential and Transmission Shafts

The differential splits the engine torque on the left and right transmission branches while allowing for different revolution speeds on the two shafts. The usual mechanical realization of this device employs epicycloidal trains.

In the case of a front engine forward traction driveline (FF layout), the differential is integrated in the gearbox just after the final reduction stage. Wheel shafts are thus directy connected to the two sides of the gearbox. FF layout is the most commonly used in consumer cars due to its compactness.

In a front engine rear traction driveline (FR layout) the gearbox in the front of the vehicle is connected through a main shaft to the differential placed between the two rear wheels. FR layout allows higher acceleration due to the load transfer on the rear wheels and is mostly used for large sedans and luxury cars.

A rear-engine rear-traction driveline (RR layout) is in principle similar to a FF layout connected to the rear shafts. An RR layout combines the traction advantage of a FR layout with a better load distribution and a lower moment of inertia at the expense of habitability. This layout is mainly limited to sports cars.

Finally, an all-wheel drive vehicle has three differentials: one for splitting torque on front and rear axles and two, one for each axle for splitting between the right and left tires.

Transmission shafts are basically steel shafts with homokinetic joints on each end to allow for wheel movements. By design, the right and left shafts have the same inertia but, due to the different length, have different stiffness coefficients.

Tires

The final element of the driveline, their radius defines, together with the gearbox reduction ratio, the total reduction ratio of the driveline. The common empirical unit of measure of this ratio in the automotive industry is the so-called $V1000$, *i.e.* the vehicle speed expressed in km/h corresponding to a 1000 rpm engine revolution speed. Excluding incidental maneuvers and other extreme cases the clutch comfort is independent of the tire's performances.

2.2 Models

2.2.1 Simulation Model

Model Structure

The detailed simulation model is composed of three parts, one main part and two auxiliary components. The main part captures the dynamic of the powertrain, while the static washer-spring and the clutch hydraulic control model transform, respectively, the throttle pedal position x_t and the clutch pedal position x_c in engine torque Γ_e and normal force F_n exerted on the friction surfaces. The relations between these last quantities are given essentially by look-up tables.

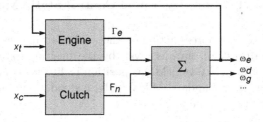

Figure 2.8. Interconnection scheme of the model's three parts

Engine Torque-generation Model

The torque-generation model is fairly simple: the throttle look-up table saturated by the maximal available torque at the current engine speed is the torque target specified by the driver through the throttle pedal.

This signal is filtered to simulate the intake pressure dynamic in indirect-injection engines; this filtering is not used for simulating Diesel direct-injection engines.

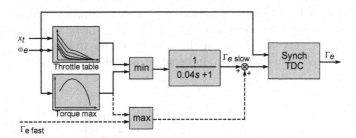

Figure 2.9. Engine torque-generation model for an indirect-injection engine. The dotted part model an eventual torque reserve use. No filtering is present in the case of a direct-injection engine.

The dotted part of the model in Figure 2.9 models a possible use of torque reserve in order to have a faster torque control that, still limited by the maximal torque available at a given engine speed, is not slowed down by the intake pressure dynamics.

To guarantee the synchronization of engine torque changes with the passage of one of the pistons through a TDC a zero-order holder is introduced having a sampling time controlled by the engine speed. The resulting signal has the same characteristics as the engine mean torque signal generated by the engine control unit.

Washer-spring and Clutch Hydraulic Control Static Model

Neglecting the centrifugal forces acting on the washer-spring the normal force F_n exerted on the pressure plate is given by the position x_b of the washer spring's fingers. For an AMT vehicle this position is directly controlled by the hydraulic actuator; in a standard MT vehicle, instead, the clutch hydraulic control relays the clutch pedal position.

In both cases the final relation is a simple monodimensional look-up table; more details on its actual determination starting from semi-static bench measures are available in [10].

Powertrain Model

Neglecting the oscillations of the engine on its suspensions the whole powertrain can be easily described as a monodimensional mechanical system as it can be seen in Figure 2.10.

From left to right we have the engine and the primary DMFW masses, the DMFW non-linear spring and viscous damper,[8] the secondary DMFW mass to which the clutch is connected. On the right side of the clutch we have the gearbox inertia calculated on the primary shaft, the reduction ratio α and the differential splitting the torque between the two transmission branches. On each side we have the transmission shaft mass, its stiffness and damping and the wheel mass. The link between the wheels and the vehicle mass is made through a lumped LuGre tire ground contact model [34].

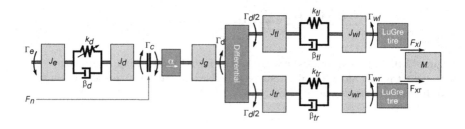

Figure 2.10. Advanced simulation model with 14 degrees of freedom

---

[8] The oscillation damping inside the DMFW is provided by the lubricated friction of the springs on the internal wall of the DMFW. The viscous damping coefficient is strictly valid only for one given engine speed due to the centrifugal force pushing the springs against the wall but this relation is usually ignored since very few measures are available.

The dynamic equations of the masses together with the three springs' Hooke's laws give

$$J_e \dot{\omega}_e = \Gamma_e - k_d(\theta_d) - \beta_d (\omega_e - \omega_d)$$
$$J_d \dot{\omega}_d = k_d(\theta_d) + \beta_d (\omega_e - \omega_d) - \Gamma_c$$
$$J_g \dot{\omega}_g = \alpha \Gamma_c - \Gamma_d$$
$$J_{tl} \dot{\omega}_{tl} = \Gamma_d/2 - k_{tl}\theta_{tl} - \beta_{tl}(\omega_{tl} - \omega_{wl})$$
$$J_{tr} \dot{\omega}_{tr} = \Gamma_d/2 - k_{tr}\theta_{tr} - \beta_{tr}(\omega_{tr} - \omega_{wr})$$
$$J_{wl} \dot{\omega}_{wl} = k_{tl}\theta_{tl} + \beta_{tl}(\omega_{tl} - \omega_{wl}) - R_w F_{xl}$$
$$J_{wr} \dot{\omega}_{wr} = k_{tr}\theta_{tr} + \beta_{tr}(\omega_{tr} - \omega_{wr}) - R_w F_{xr}$$
$$\dot{\theta}_d = \omega_e - \omega_d$$
$$\dot{\theta}_{tl} = \omega_{tl} - \omega_{wl}$$
$$\dot{\theta}_{tr} = \omega_{tr} - \omega_{wr}$$
$$M\dot{v} = F_{xl} + F_{xr}.$$

All the symbols employed in these equations are defined in Table 2.1.

Table 2.1. Definition of the symbols used in (2.1a)

J_e	engine inertia	J_d	DMFW secondary inertia
J_g	gearbox inertia	J_{tl},J_{tr}	left, right trans. inertias
J_{wl},J_{wr}	left, right wheel inertias	M	vehicle mass
ω_e	engine speed	ω_d	DMFW speed
ω_g	gearbox speed	ω_{tl},ω_{tr}	left, right trans. speeds
ω_{wl},ω_{wr}	left, right wheel speeds	θ_d	DMFW torsion
θ_{tl},θ_{tr}	left, right trans. torsion	$k_d(\theta_d)$	DMFW non-linear stiffness
β_d	DMFW damping	k_{tl},k_{tr}	left, right trans. stiffness
β_{tl},β_{tr}	left, right trans. damping	R_w	wheel radius
v	vehicle speed	α	gearbox ratio
Γ_e	engine torque	Γ_c	clutch torque
Γ_d	differential torque	F_{xl},F_{xr}	left, right tang. forces

The clutch torque is given by a LuGre friction model, which provides a continuous, albeit non-linear, model instead of the usual linear hybrid models found in the literature. For a detailed description of the model and a physical explanation of its parameters please see Appendix C. The resulting equations are

$$\dot{z}_c = \omega_d - \alpha\omega_g - \sigma_{0c}\frac{|\omega_d - \alpha\omega_g|}{g_c(\omega_d - \alpha\omega_g)}z_c \tag{2.1a}$$

$$g_c(\omega_d - \alpha\omega_g) = \alpha_0 + \alpha_1 e^{-\left(\frac{\omega_d - \alpha\omega_g}{\omega_{0c}}\right)^2} \tag{2.1b}$$

$$\Gamma_c = F_n \left[\sigma_{0c}z_c + \sigma_{1c}e^{\left(\frac{\omega_d - \alpha\omega_g}{\omega_{dc}}\right)^2}\dot{z}_c + \sigma_{2c}(\omega_d - \alpha\omega_g)\right]. \tag{2.1c}$$

The tire longitudinal forces due to the wheels contact with the ground are defined through an average lumped LuGre tire ground contact model [34]

$$v_{ri} = v - R_w \omega_{wi}$$

$$g_i(v_{ri}) = \mu_{ci} + (\mu_{if} - \mu_{ci})e^{-|v_{ri}/v_{si}|^{1/2}}$$

$$\dot{z}_i = v_{ri} - \sigma_{0i}\frac{|v_{ri}|}{g_i(v_{ri})}z_i - \kappa|\omega_{wi}R_w|z_i$$

$$F_{xi} = F_z\left[\sigma_{0i}z_i + \sigma_{1i}\dot{z}_l + \sigma_{2i}(v_{ri})\right].$$

with $i = \{r, l\}$. The κ parameter, absent in the point contact friction model used for the clutch, captures the distributed nature of the tire contact. The most prominent difference induced by this parameter is a v_{ri} continuous steady state friction force for $\omega_{wi} \neq 0$. For more information on tire friction dynamics and models please see [7].

The differential torque Γ_d is derived from the following relation

$$\omega_g = 1/2(\omega_{tr} + \omega_{tl})$$

and its time derivative.

Model Validation

In order to validate the driveline model a test campaign has been effectuated with a Megane II 2.0 gasoline (F4R) test vehicle. The engine speed, vehicle speed, engine torque target and clutch pedal position have been recorded and fed to the driveline model and clutch static model to compare results. Since the characteristic curve of the clutch on the car was not available, static bench measures of another clutch of the same model have been used thus introducing some error.

Figure 2.11 shows the results of a standing-start simulation alongside the corresponding measures. The simulated vehicle speed is quite close to the measures; the error on the engine speed during the last part of the engagement is probably due to a wrong estimation of the engine torque.

Figure 2.12 shows the results of a simulated 1-2 gearshift together with the actual measures. Despite a slight timing error, the driveline oscillation induced by the synchronization in the simulation is quite similar to the one actually measured on the vehicle.

These results show that even if the driveline model is not precise enough to give an exact simulation of the driveline behavior, it can nonetheless provide a good estimation of the comfort performances.

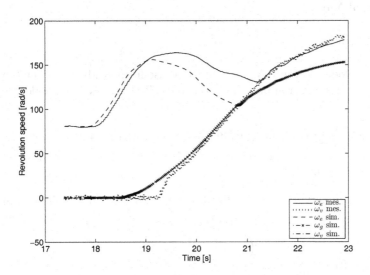

Figure 2.11. Driveline model validation for a standing-start of a Megane II equipped with a 2.0 gasoline engine (F4R)

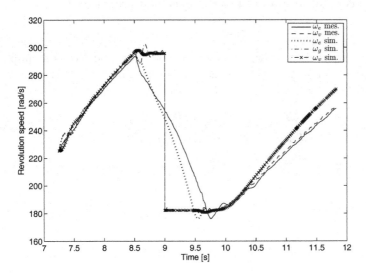

Figure 2.12. Driveline model validation for a 1-2 gearshift on a Megane II equipped with a 2.0 gasoline engine (F4R)

2.2.2 Control Model

Driveline Model

The previous model is way too complex to be useful in studying the clutch comfort and to design an appropriate controller. To simplify this model we assume that:

- the two branches of the driveline are perfectly symmetric; and
- the tires have a perfect adherence and no transitory effects on tire ground contact are present.

Assuming the symmetry of the driveline the two branches can be collapsed in one having

$$J_t = J_{tr} + J_{tl}$$
$$k_t = k_{tr} + k_{tl}$$
$$\beta_t = \beta_{tr} + \beta_{tl}$$
$$J_w = J_{wr} + J_{wl}.$$

Thanks to the second simplifying hypothesis the vehicle mass can be simply written as an equivalent rotational inertia plus the wheel inertias

$$J_v = Mr_w^2 + J_w.$$

The driveline downstream of the gearbox can be thus modeled as a simple linear spring damper system whose behavior can be expressed relative to the gearbox primary shaft

$$J_g' \dot{\omega}_g' = \Gamma_c - k_t' \theta_t' - \beta_t' \left(\omega_e' - \omega_g' \right)$$
$$J_v' \dot{\omega}_v' = k_t' \theta' + \beta_t' \left(\omega_e' - \omega_g' \right)$$
$$\dot{\theta}' = \omega_e' - \omega_g',$$

where

$$J_g' = \frac{J_g + J_t}{\alpha^2}$$
$$k_t' = \frac{k_t}{\alpha^2}$$
$$\beta_t' = \frac{\beta_t}{\alpha^2}$$
$$J_v' = \frac{J_v}{\alpha^2}.$$

In first or second gear the poles induced by the transmission stiffness are largely dominant relative to the poles due to the DMFW springs meaning that the uncomfortable oscillations that are the subject of this thesis are mainly due to the transmission torsion.

We can neglect, therefore, the DMFW stiffness and add the secondary DMFW disk mass to the engine mass

$$J'_e = J_e + J_d$$

We have, finally, a very simple driveline model, having just four state variables:

$$J'_e \dot{\omega}'_e = \Gamma_e - \Gamma_c \tag{2.2a}$$
$$J'_g \dot{\omega}'_g = \Gamma_c - k'_t \theta'_t - \beta'_t \left(\omega'_e - \omega'_g\right) \tag{2.2b}$$
$$J'_v \dot{\omega}'_v = k'_t \theta' + \beta'_t \left(\omega'_e - \omega'_g\right) \tag{2.2c}$$
$$\dot{\theta}' = \omega'_e - \omega'_g. \tag{2.2d}$$

The relation between the previous model parameters and those of the simplified model is:

$$J'_e = J_e + J_d$$
$$J'_g = \frac{J_g + J_{tr} + J_{tl}}{\alpha^2}$$
$$k'_t = \frac{k_{rt} + k_{lt}}{\alpha^2}$$
$$\beta'_t = \frac{\beta_{rt} + \beta_{lt}}{\alpha^2}$$
$$J'_v = \frac{1}{\alpha^2} \left(r_w^2 M + J_{wr} + J_{wl}\right).$$

This model captures the essential part of the dynamic behavior of the driveline as can be seen in Figure 2.14.

Considering a constant sliding speed the LuGre friction model 2.1 gives

$$\Gamma_c = g_c(\omega'_e - \omega'_g)\text{sign}\left(\omega'_e - \omega'_g\right) F_n. \tag{2.4}$$

Since the surface stiffness, modeled by the σ_0 parameter of the LuGre model, is very high, during the sliding phase, but for the very last few instants, the internal dynamic of the model is much faster than the variations of the sliding speed. The global behaviour of the friction model can be, thus, assimilated to a simple Coulomb friction model

$$\Gamma_c = 2\mu_d r_c F_n \text{sign}\left(\omega'_e - \omega'_g\right), \tag{2.5}$$

where μ_d is the Coulomb friction coefficient, r_c the clutch friction pads mean radius. The constant 2 is due to the double friction interaction flywheel friction disk and friction disk pressure plate.

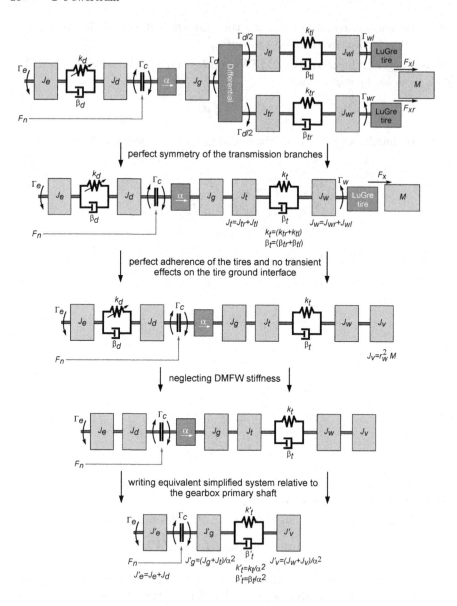

Figure 2.13. Step-by-step derivation of the simplified model from the complete driveline model.

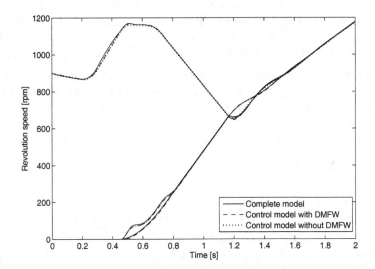

Figure 2.14. Standing-start simulations on flat ground using the same pedal position profiles and a complete driveline model, a simplified model including a liberalized DMFW and a simplified model without DMFW. Models parameters represent a 2.0l gasoline (F4R) Megane II.

During a normal clutch engagement, *i.e.* a comfortable one, the sliding speed doesn't change sign. This observation allows a further simplification:

$$\Gamma_c = 2\mu_d r_c F_n = \gamma F_n \tag{2.6}$$

for standing-starts and upward gearshifts and

$$\Gamma_c = -2\mu_d r_c F_n = -\gamma F_n \tag{2.7}$$

for downward gearshifts.

Clutch Actuator Model

In an AMT or clutch-by-wire architecture, the washer-spring's fingers position x_b is directly controlled by a hydraulic actuator. The higher level engagement strategies that are responsible for the comfort do not specify directly this position but rather for a required level of transmitted torque $\bar{\Gamma}_c$. This target is translated by the low-level routines in a clutch's finger position through the inversion of the estimated $\Gamma_c(x_b)$ characteristic. This curve is learned and updated through least square estimation of the parameters of a third-order polynomial.

During the design of the engagement control strategies presented in the following chapters we will first assume a perfect estimation of the curve and an infinite actuator dynamics. These two hypotheses coupled with a positive sliding speed allow the clutch to be considered as a simple torque actuator.

In order to improve the robustness of the engagement strategy a supplementary corrective multiplicative factor has been added to the estimation of the clutch characteristic. This value is obtained through the use of a friction-coefficient observer or a clutch-torque observer presented in Chapter 5.

2.2.3 Driver Model

In order to successfully simulate a standing-start or a gearshift two finite state machines reproducing the driver's behavior have been introduced.

If we consider a standing-start the initial condition is a standing vehicle, first gear engaged, clutch completely open and no throttle. The finite state machine, showed in Figure 2.15, goes through the following steps in order to complete a standing-start:

- *A: Reaching the Contact Point* - the clutch is rapidly closed till the pressure plate makes contact with the friction and the vehicle starts to move. The throttle pedal is lightly pressed;

- *B: Obtaining a Given Level of Acceleration (First Part)* - the closing of the clutch proceeds at a slower rate with the throttle pedal lightly pressed till the engine speed starts to drop due to the increase of the clutch torque;

- *C: Obtaining a Given Level of Acceleration (Second Part)* - in reaction to the drop of the engine speed the throttle pedal is pressed further, while the closing of the clutch proceeds till the required acceleration level is reached;

- *D: Wait* - once the required acceleration level is attained the position of the two pedals is kept constant till the engagement is over; and

- *E: Final Closing of the Clutch* - after the synchronization the clutch is completely closed; the throttle pedal might be further pressed to accelerate the vehicle. When the clutch is fully closed the standing-start procedure ends.

In the case of an upward gearshift the initial condition is an accelerating vehicle under the impulsion of the engine torque with a clutch fully engaged. The finite state machine, showed in Figure 2.16, goes through the following steps in order to complete a standing-start:

- *A: Reaching an Engine Speed Target* - under the engine torque the engine and the vehicle are accelerated till a threshold level is met, triggering the gearshift;

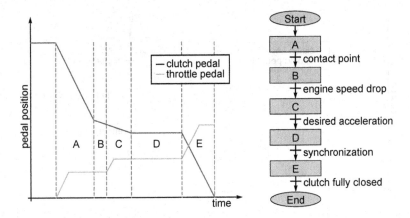

Figure 2.15. Finite state machine generating the throttle and clutch pedal positions for a standing-start.

- *B: Disengagement of the Clutch* - the clutch is rapidly disengagedn thus opening the driveline; to avoid an excessive engine speed the throttle pedal is also released;

- *C: Gear Selection and Engagement* - once the driveline is open the old gear is disengaged and the new one is selected and engaged;

- *D: Synchronization* - due to the gearshift the revolution speeds of the crankshaft and the gearbox primary shaft are different, the clutch is progressively closed to synchronize the two shafts; and

- *E: Acceleration* - once the shafts are synchronized and the clutch fully engaged the throttle pedal is again pressed.

This sequence applies an engine torque only after the shafts are fully synchronized. In this case, called throttle-less engagement, is not the most common since usually the driver presses the throttle pedal before while closing the clutch but is the worst case concerning the driveline oscillations.

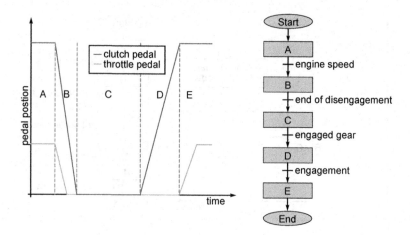

Figure 2.16. Finite state machine generating the throttle and clutch pedal positions for a gearshift.

3

Clutch Comfort

3.1 Detailed Analysis of the Clutch Use

3.1.1 When the Clutch Is Used

On an MT or AMT vehicle the clutch is used in three specific cases:

- *Standing-start* As noted earlier, the engine cannot sustain a revolution speed lower than the idle speed. During a standing-start the clutch allows for a smooth acceleration of the vehicle from zero to cruising speed.

- *Creeping* The creeping mode assures a very slow vehicle movement through a controlled slipping speed of the clutch. This mode, normally used while parking the vehicle, does not introduce any specific comfort criteria and thus is not further analyzed in this book.

- *Gear Shifting* A gearshift induces a speed difference between the crankshaft and the gearbox primary shaft. The engagement of the clutch assures the synchronization of the two shafts.

Clutch comfort is a wide subject straddling different competence fields such as acoustics, ergonomics and driving pleasure and is subject to strict cost and endurance constraints.

The clutch has an impact on vehicle acoustics and vibration performances since the MT clutch hydraulic circuit is a direct link between the engine and the cockpit. To avoid a transmission of the engine vibrations to the clutch pedal hydraulic filters are placed on the hydraulic line connecting the CMC to the CSC. The clutch itself can also be a source of vibrations either for purely mechanical reasons, like a bent friction disk, or because of a more complex thermo-elastic-induced oscillation, called *hot judder*, first highlighted by Barber [3] and well studied since [2, 35].

The clutch pedal of an MT vehicle is part of the control interface used by the driver and, thus, is subject to ergonomics criteria, such as the pedal stroke length, the maximal force needed to operate the clutch and the position on the pedal stroke at which the pressure plate makes contact with the friction disk. The first two elements are deeply related to the driving position given by the global architecture of the vehicle. The contact point, instead, should have a fixed optimal position around 60% of the pedal stroke.

Finally the behavior of the clutch itself when used by the driver introduces comfort criteria which are somewhat more difficult to quantify but are an important component of the driving pleasure. A first element of these comfort criteria, valid only for MT vehicles, is the so-called clutch disability, *i.e.* the ease with that the driver can attain the desired transmitted torque. The second element, closely related to the first but valid for both MT and AMT vehicles, is the clutch ability of assuring a lurch-free engagement. The rest of this work will focus mainly on this aspect of the clutch comfort that is not only very important for the driving pleasure but contributes to the perceived quality of the vehicle.

In order to better define the study subject, the use of the clutch during a standing-start and an upward gearshift is analyzed in detail. A downward gearshift has roughly the same structure as an upward shift but for some changes in the slipping speed and torque direction. As previously noted, since the creeping mode does not introduce any specific comfort criteria it is not further investigated.[1]

3.1.2 Standing-start Analysis

In principle, a standing-start maneuver is extremely simple: it just consists in a gradual closure of the clutch while pressing the throttle pedal. Synchronization between the two actions allows to smoothly accelerate the vehicle while limiting the engine speed peak value.

In an MT vehicle, where the driver directly controls the clutch, the closure can be divided into four phases:

- a fast rise of the pedal till the contact point is reached, *i.e.* the pressure plate makes contact with the friction disk;
- a slower closing action to increase the vehicle. acceleration to reach the desired acceleration level;
- a constant position till the end of the sliding phase; and
- a final fast closure.

---

[1] AMT vehicles, in order to simulate an AT behavior, start creeping as soon as the brakes are released. This behavior does not affect the standing-start analysis exposed in the following sections and is therefore ignored.

AMT strategies are usually programmed to have a similar behavior even if sometimes, in order to reduce the synchronization-induced oscillations and avoid an excessive engine speed drop, the clutch torque is limited.

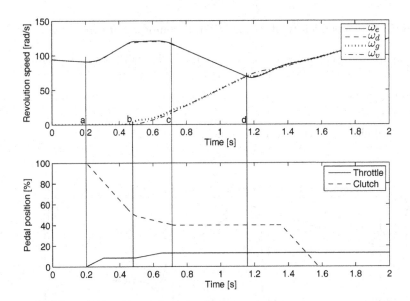

Figure 3.1. Simulation of a standing-start on flat ground. a beginning of the standing-start b contact point c desired acceleration level is reached d synchronization.

Figure 3.1 shows the results of a simulation run of a standing-start on flat ground of an MT vehicle. At the beginning of the simulation the engine is idling, the vehicle is standing and the first gear is engaged with a fully open clutch. The a point marks the beginning of the standing-start with a slight depression of the throttle pedal and a fast clutch closure till the contact point is reached at b. The engine accelerated by the growing engine torque and not yet braked by the clutch torque revs up. Once the contact point is reached the vehicle begins moving forward under the effect of an increasing clutch torque. When the clutch torque becomes higher than the engine torque the engine speed starts to drop. As a reaction the driver pushes further the throttle pedal while closing the clutch till the desired level of acceleration is reached (point c). After this point the driver simply waits without any further change in the pedal positions for the synchronization to happen, after which the clutch is completely closed. When the crankshaft and the gearbox primary shaft are synchronized (point d) the clutch behaves like a simple linking element inducing a sudden change in transmitted torque that causes a highly uncomfortable oscillation of the driveline visible in Figure 3.1 just after point d. The causes of this oscillation and its effects on the perceived comfort will be detailed in the following sections.

3.1.3 Upward Gearshift Analysis

A gearshift, independently of its direction, is composed of three phases:

- a complete disengagement of the clutch coupled with a sudden reduction of the engine torque to avoid the engine to revving up;
- disengagement of the current gear followed by the selection and engagement of the new gear; and
- gradual closure of the clutch to synchronize the two shafts.

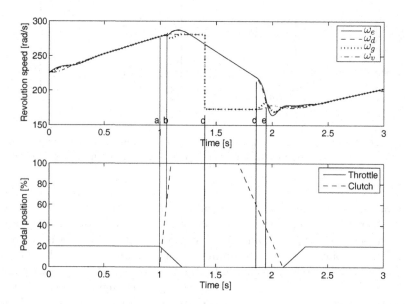

Figure 3.2. Simulation of a throttle-less 1-2 upward gearshift. *a* beginning of the gearshift *b* clutch disengagement *c* new gear engagement *d* beginning of the slipping phase *e* synchronization.

These three phases can be easily seen in Figure 3.2, showing the results of a simulation of an upward 1-2 gearshift for an MT vehicle. At the beginning of the simulation the clutch is fully engaged, the first gear is engaged and the vehicle is accelerated by the engine. The *a* point marks the beginning of the gearshift characterized by a simultaneous rapid disengagement of the clutch and a raising of the throttle pedal. Due to the action on the clutch pedal the normal force squeezing the clutch disk is reduced till its complete disengagement in b. Since the engine becomes free of any load while the intake pressure has not yet dropped the engine speed has a slight increase before dipping under the effect of the negative torque due to the internal friction and pumping work. While the vehicle is coasting due to its inertia, the engagement of the second gear induces a sudden change in the speed of the gearbox primary shaft (point c).[2] The closure of the clutch during the third and final phase synchronizes the crankshaft and the primary shaft speeds and ends the gearshift operation. This engagement, between d and e in the figure, is quite fast with a sliding phase of just some tenths of a second. The clutch torque during this phase can be quite large and induces uncomfortable oscillations in the driveline after the two shaft speeds are synchronized in point e as in the case of a standing-start.

3.1.4 Clutch Torque at Synchronization

During the sliding phase the clutch torque

$$\Gamma_c = 2\mu_d r_c F_n \tag{3.1}$$

is controlled by the normal force F_n acting on the friction surfaces. After the synchronization instant t_s the clutch disk adheres to the flywheel and the clutch behaves like a simple connecting rod.

If we neglect the transmission torsion the driveline simplified model (Equation 2.2) is reduced to two masses J'_e and $J_1 = J'_g + J'_v$ connected by the clutch (Figure 3.3).

Figure 3.3. Synchronization analysis under a perfectly rigid transmission assumption

---

[2] This speed change is not instantaneous, the primary shaft is brought to the correct speed by the action of the synchronization meshes. While being of paramount importance for the gearbox comfort, this dynamic is irrelevant for the clutch comfort and thus is ignored in this analysis.

In t_s the clutch torque jumps from the value given by (Equation 3.1) to

$$\Gamma_c(t_s^+) = \frac{J_1}{J_e' + J_1}\Gamma_e.$$ (3.2)

In the case of a finite transmission stiffness the synchronization constraint, i.e. $\omega_e' = \omega_g'$ and $\dot\omega_e' = \dot\omega_g'$, imposed on the Equations 2.2 gives clutch torque oscillating around the equilibrium value

$$\Gamma_{ceq} = \frac{J_g' + J_v'}{J_e' + J_g' + J_v'}\Gamma_e,$$ (3.3)

with an initial value of

$$\Gamma_c(t_s^+) = \frac{J_g'}{J_e' + J_g'}\Gamma_e + \frac{J_e'}{J_e' + J_g'}\left(k_t'\theta' + \beta_t'\left(\omega_g' - \omega_v'\right)\right),$$ (3.4)

as shown in Figure 3.4.

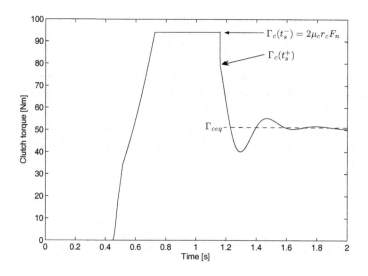

Figure 3.4. Clutch torque during a simulated standing-start

3.1.5 Clutch-related Driving Comfort

MT and AMT vehicles share the evaluation criteria used to measure the comfort of a standing-start or gearshift operation.

For a standing-start operation the factors influencing the driver's subjective comfort perception are:

- overall duration of the operation (clutch slipping time);
- ease with which a torque target is met; and
- oscillations of the driveline after synchronization (lurch).

In the case of a gearshift the driver has no acceleration target and only the first and third criteria are valid. Taken to the extreme an exceptionally comfortable clutch for gear-shifting would allow a very fast engagement without any driveline oscillation in the aftermath of the synchronization.

The ease with which a torque target is met, particularly important for an MT vehicle, is strictly linked to the ergonomics of the control that is basically given by the $\Gamma_c(x_b)$ characteristic determined by the washer spring and flat spring stiffness.

The remaining two factors are somewhat autonomic since in order to reduce the slipping time one has to increase the clutch torque inducing stronger oscillations in the driveline after the synchronization. The need for a compromise between the duration of the operation and the lurch level induces an important difference between MT and AMT comfort standards. In the former case, in fact, the compromise is, more or less consciously, chosen by the driver while in the latter case the gearbox control unit makes the decision. An AMT driver is much more sensitive to the oscillations and the overall comfort of a standing-start or gearshift operation since he has no control whatsoever over the clutch behavior for a given throttle position. This explains, for example, why a total upward gearshift operation duration of one second is normal on an MT vehicle but excessively long on an identical AMT vehicle.

3.2 Influence of the Driveline Parameters

The driveline physical parameters have a strong influence on a vehicle lurch performance. Three main trends, namely the increase of clutch transmissible torque, the transmission stiffness reduction and the driveline's parasite friction reduction make this performance more and more critical.

In order to avoid any unwanted slippage the maximal torque the clutch can handle must be greater than the engine peak output torque times a safety factor. The peak torque of Diesel turbocharged engines, thanks to technical evolution, has greatly increased; the last twelve years have seen a 322% increment in output peak torque for the same engine displacement in Renault Diesel engines, as shown in Table 3.1. The use of higher-capacity clutches induces a loss in the driver-perceived ability to control the transmitted torque, called dosability, and an increase in the effort needed to activate the clutch pedal.

To better filter out the engine acyclicity and thus improve the noise, vibration and harshness (NVH) performances of the vehicle the transmission-shaft

Table 3.1. Output peak torque evolution of Renault Diesel engines

Comm. name	Tech. name	Year	Γ_e max	Technology
1.9 D	F8Q	1994	$118Nm$	indirect inj.
1.9 DT	F8Qt	1996	175 Nm	indirect inj. turbo
1.9 dTi	F9Q	1997	160 Nm	direct inj. turbo
1.9 dCi	F9Q	1999	200 Nm	common rail turbo
1.9 dCi	F9Q	2000	250 Nm	common rail turbo
1.9 dCi	F9Q	2002	300 Nm	common rail turbo
2.0 dCi	M9Rb	2006	400 Nm	common rail turbo

stiffness is reduced. Given the same amplitude a lower oscillation frequency is more uncomfortable since the human sensibility to vibrations is higher at lower frequencies.

Finally, the minimization of parasite friction losses of the driveline for the fuel efficiency sake, reduces the damping coefficient thus multiplying the number of oscillations perceived by the driver.

3.3 State-of-the-art

3.3.1 Manual Transmission

In the actual state-of-the-art the only way of improving the clutch comfort performances in an MT vehicle is to operate an intelligent choice of the driveline's physical parameters coupled with an active oscillation damping through the control of the engine torque.

The restriction of the possible choices due to the trends previously highlighted and an improvement of comfort standards motivates the research for an alternative solution.

3.3.2 Automated Manual Transmission

The availability of simultaneous engine and clutch torque control allows for the introduction of advanced lurch-avoidance strategies.

The problem of an AMT vehicle engagement control has been widely analyzed in the literature. A recent compilation [19] of the articles concerning the clutch modeling and control lists more than 27 contributions published between 1995 and 2006. Several different approaches to the problem have been proposed: quantitative feedback theory (QFT) [32], fuzzy logic [33] and [31], model predictive control (MPC) [5], decoupling control [14] and, finally, optimal control [16] and [15].

The last two cited articles introduce a synchronization criterion called the *GV no-lurch condition* aiming for the continuity of the time derivative of the clutch speed over the synchronization instant t_s for a driveline model equivalent to (Equation 2.2) and under the hypothesis of continuity of the engine speed and transmission torsion. Since, after the synchronization, we have $\omega_e' = \omega_g'$ we have from Equations 2.2a and 2.2b

$$\dot{\omega}_e' = \dot{\omega}_g' = \frac{1}{J_e' + J_g'} \left(\Gamma_e - k_t' \theta_t - \beta_t' \left(\omega_g' - \omega_v' \right) \right). \tag{3.5}$$

Imposing the continuity condition $\dot{\omega}_g'(t_s^+) - \dot{\omega}_g'(t_s^-)$ we have from (2.2b) and (3.5)

$$\Gamma_c(t_s) = \frac{J_g'}{J_e' + J_g'} \Gamma_e(t_s), \tag{3.6}$$

which expressed as a function of the sliding speed $\omega_s = \omega_e' - \omega_g'$, following the original formulation of the cited articles, is equivalent to

$$\dot{\omega}_s(t_s) = \frac{\mathrm{d}}{\mathrm{d}t} \left(\omega_e' - \omega_g' \right)(t_s) = \frac{1}{J_e'} \Gamma_e(t_s) - \frac{J_e' + J_g'}{J_e' J_g'} \Gamma_c(t_s) = 0. \tag{3.7}$$

Unfortunately few of these methods have been implemented on actual prototypes and the actual strategies found on stock cars are just hand-tuned look-up tables aiming at having a clutch torque lower than the engine torque. This choice, inducing a low oscillation level but a long slipping time, is partially motivated by the commercial positioning of the AMT as a cheap replacement of an automatic transmission.

A more sportive image of the AMT due to the introduction of a gearshift control on the wheel and the limitation of the performances available on an MT vehicle justify the study of more advanced strategies.

3.3.3 Clutchless Gearshifting

In heavier vehicles like trucks an alternative clutchless gearshifting strategy for AMT can found. The basic idea is to control the engine torque in order to realize a virtual clutch [29, 30]. In such a scheme the torque on the transmission is first brought to zero to allow the disengagement of the first gear before assuring the speed synchronization through the control of the engine torque before engaging the second gear thus emulating the behavior of an open and sliding clutch.

Even if the clutchless gearshifting strategy is not adapted for smaller vehicles where the smaller inertias are not a serious problem for the clutch's friction pad life, the observer-based active oscillation damping implemented in this controller can be an interesting improvement of the standard active damping strategies based on the engine-speed oscillations.

3.4 Motivation and Methodology

3.4.1 A Manual Transmission in Troubled Waters

As previously highlighted the technical evolution of the driveline elements and the new challenges that it has to face makes it increasingly difficult to assure a correct level of comfort for the clutch, specially in the case of an MT architecture.

In order to better understand the problem and find innovative solutions a research project has been started in collaboration between the Driving Comfort department in Renault and the Laboratoire d'Automatique de Grenoble; this book details the main results of this effort.

After a first exploratory phase whose main objective was the understanding and modeling of the different powertrain and driveline elements, attention has been devoted to the mechanical passive means of improving the clutch comfort. As a consequence of this first study, whose results are given in detail in following part of this chapter, the introduction of an active element on the hydraulic control of the clutch allows the partial decoupling of the washer spring's finger position from the clutch pedal position.

3.4.2 Passive Means of Increasing the Clutch Comfort

A passive mechanical or hydraulic mean of increasing the clutch comfort is very attractive due to its inherent simplicity, endurance and low cost. Since the design of the clutch itself is subject to many hard constraints most of the attention has been devoted to the hydraulic circuit connecting the clutch pedal to the CSC.

Two solutions have been studied: a filtering action on the hydraulic circuit, similar in conception to what is already done to avoid the engine vibrations to reach the clutch pedal but at a much lower frequency, and a variable reduction system aiming to maximize the dosability.

Hydraulic Filtering

The basic idea behind the introduction of a filter on the hydraulic circuit is to improve the dosability by limiting the effects of excessively rapid pedal movements.

Two means of intervention have been investigated: a low-pass filter, similar in conception to a very soft engine vibration filter, and a clutch pedal speed limiter. A standing-start operation has been chosen as the test bed since it requires a finer control of the clutch torque compared to a gearshift.

Two criteria have been used to quantify the filtering action effects: a dynamic quadratic cost function and a static criterion on the driveline state at synchronization. The first

$$J_{dyn} = \int_{t_0}^{t_s} \left(a(\omega_e - \omega_g)^2 + b(\omega_g - \omega_v)^2 + cu^2 \right) dt, \qquad (3.8)$$

where $u = \mathrm{d}/\mathrm{d}t\Gamma_e$, weights the square of the clutch sliding speed, of the torsional speed of the transmission shafts and of the actuator speed. The second criterion is simply the distance in quadratic norm between the driveline state vector at the synchronization instant and the corresponding vector of ideal synchronization condition defined by Equations 4.13, 4.10, 4.12, and 4.11. To simplify result interpretation the output values have been normalized with respect to the results corresponding to a minimal influence of the filtering action.

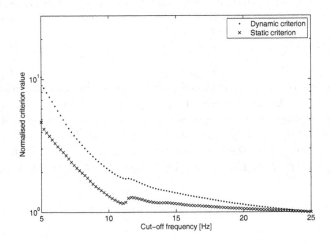

Figure 3.5. Evolution of the static and dynamic criteria for the low-pass filtering action as a function of the cut-off frequency. Results have been normalized with respect to the results obtained for a 25 Hz cut-off frequency.

Low-pass Filter

This kind of filtering action would probably be achieved through the use of a filter similar to that used to prevent engine vibrations to reach the clutch pedal; the actual mechanical implementation of the filter has not been taken into consideration due to the poor performances of this solution. A low-pass filter with a low cut-off frequency induces an important delay between the pedal position and the actual reaction of the vehicle to this position. This

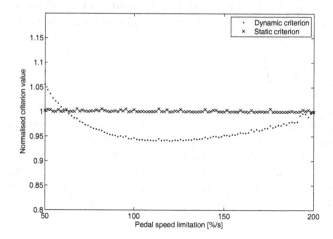

Figure 3.6. Evolution of the static and dynamic criteria for the pedal speed limiter as a function of the speed limit. Results have been normalized with respect to the results obtained for a 200%/s limit speed.

delay, unsurprisingly, makes the control of the transmitted clutch torque more difficult and neutralizes any eventual positive effect due to the smoothing of the driver's input. As clearly shown by the results of the two criteria in Figure 3.5, a low-pass filter actually reduces the clutch comfort during a standing-start instead of improving it. Since the closure speed, after a small transition phase, is not affected by the low-pass filter, we can safely conclude that the clutch comfort for a gearshift is not affected by the introduction of a low-pass filter.

Speed Limiter

The clutch pedal is pushed, through the hydraulic system, by the washer-spring's fingers; a restriction in the section of the hydraulic system dampens this movement and limits the maximum speed the pedal reaches if suddenly released. Since the speed limiter does not introduce any delay, it does not adversely affect the clutch torque control ability of the driver. The results of the static criterion shown in Figure 3.6 show clearly that the oscillation level, *i.e.* the distance from the ideal synchronization conditions, is independent of the pedal speed. This result, apparently in opposition to the experimental evidence showing a strong connection between the pedal speed and the oscillation level, is due to the fact that the driver model does not link the speed of the movement with the desired acceleration level as a human driver unconsciously does. The slight changes in the dynamic criterion are due to the longer sliding time for low limit speeds and a higher derivative of the clutch torque for high speeds. Similar studies involving the gearshift operation have shown, instead, a remarkable positive influence of the clutch pedal speed limitation on the clutch comfort. Unfortunately, too strong a limitation can induce a very unpleasant feeling if the pedal does not stay in contact with the driver's foot during a very fast release. For this reason the pedal speed is currently limited to 500%/s, a value only reached during extremely sportive driving or accidental maneuvers.

Variable Reduction System

Empirical evidence and experience show that a clutch having a transmissible torque curve with a contact point around 60% of the pedal stroke, a very progressive attack and not too steep an initial linear part assures a very good level of driving comfort, *i.e.* has a good level of dosability.

The shape of this curve is given by the interaction between the washer spring and the clutch disk's flat spring. The washer spring is mainly designed to satisfy the required torque capacity and to minimize the force needed to open the clutch. Size and inertia limitations, on the other hand, the possible non-linearity of the flat spring thus limiting the control over the transmissible torque curve.

An hydraulic control circuit with a variable reduction can modify this characteristic curve in order to assure a contact point around 60% of the pedal stroke and maximize the clutch dosability by reducing the pedal stroke range corresponding to a high level of transmitted torque. An example of this principle is given in Figure 3.7.

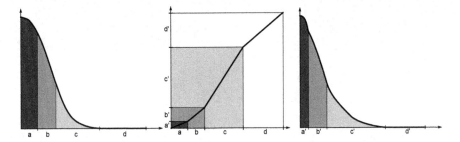

Figure 3.7. Example of a variable reduction hydraulic control system; the high torque range has been reduced in favor of the initial part improving the dosability while assuring a contact point around 60% of the pedal stroke

The main problem of this solution is the change on the force the driver has to exert on the pedal to operate the clutch. This change is a direct effect of the different reduction values. In order for the system to be ergonomically feasible an active effort compensation is needed.

3.4.3 Conclusion on the Passive Means of Improvement

The analysis of the passive means to improve the clutch comfort has shown the intrinsic limitations of this approach. The introduction of a filtering action on the hydraulic circuit can adversely influence the capability of the driver to control the clutch torque or induce an unpleasant feeling in the interaction with the clutch pedal. The variable reduction system, on the other hand, is an interesting solution allowing to better satisfy the various constraints while improving comfort by introducing a supplementary degree of freedom. Unluckily, its practical realization is quite difficult due to the large manufacturing dispersion of clutch characteristics on the one hand and the gradual change due to wear and heat of these characteristics over time for the same clutch on the other. The introduction of this kind of system including an effort-reduction system is also economically difficult to justify since its cost would be similar to a that of a clutch-by-wire or AMT system that offer a much higher potential for comfort improvement.

The main focus of this research effort, therefore, has been shifted to the control strategies of an actuated clutch. Chapter 4 presents a control strategy called *synchronization assistance* that can be implemented either on an MT or AMT architecture. The main principle of this approach is to leave the driver a complete freedom of action and to assure a perfect lurch-free synchronization by controlling the clutch torque in the very last few instants of the engagement. Chapter 5 is an extension of the previous strategy aiming at controlling an AMT system from the very beginning of the engagement. Finally, the Chapter 6 details the friction coefficient and clutch-torque observers developed for the implementation of the synchronization assistance on a Clio II AMT prototype car detailed in the Chapter 7.

Part II

Dry Clutch Engagement Control

4

Synchronization Assistance

4.1 Principle

The analysis of the passive means of improving the clutch comfort for an MT architecture has shown the need for introducing an element allowing for an active control of the clutch.

The paramount idea behind this chapter has its origin in the observation that the perceived clutch comfort is mainly affected by two elements: the total length of the engagement and the amplitude of the driveline oscillations following the synchronization. These two elements are physically interdependent: a shorter engagement time usually implies a higher oscillation level since the torque transmitted by the clutch must be increased.

The proposed solution, called *synchronization assistance*, aims to take advantage of the decoupling between the clutch pedal and the CSC position made possible by the introduction of an active element on the clutch hydraulic control system, in order to reduce to the lowest possible level the lurch at the synchronization by controlling the clutch torque in the last few instants of the engagement. This solution is completely transparent to the driver and assures an exceptional level of comfort by coupling a short engagement time with a very low lurch level.

The lurch minimization can be thought of as a *rendezvous* problem on the ω'_e and ω'_g speeds relative to the simplified control model. The GV no-lurch condition introduced by Glielmo and Vasca [14], imposing a zero time derivative of the clutch sliding speed $\omega'_e - \omega'_g$ at synchronization, can be shown to be equivalent to imposing the constraint (3.7) on the clutch torque Γ_c value relative to the engine torque Γ_e at synchronization instant t_s.

During a standard standing-start on an MT vehicle at t_s we have $\Gamma_c > \Gamma_e$, engine speed is dropping ($\dot{\omega}_e < 0$) and the vehicle is accelerating ($\dot{\omega}'_g > 0$) as can be seen in Figure 4.2.

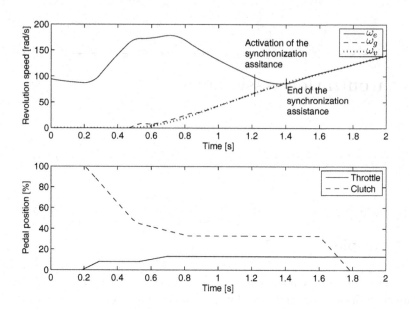

Figure 4.1. Standing-start simulation with a synchronization assistance strategy

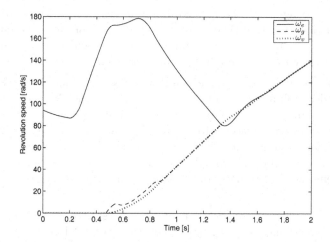

Figure 4.2. Simulation of a standard standing-start for an MT vehicle on flat ground

In an AMT vehicle the clutch torque, completely controlled by the gearbox control unit, could have been limited in order to prevent an uncomfortable lurch at the synchronization. This simplistic strategy, often used in marketed vehicles, induces a long engagement time giving the negative impression of driving a sluggish car. For this reason we will consider that $\Gamma_c > \Gamma_e$ in the final phases of the engagement independently of the transmission architecture.

To satisfy the GV no-lurch condition starting from a situation where $\Gamma_c > \Gamma_e$ either we ask the engine control unit to increase the engine torque or we partially reopen the clutch to reduce the transmitted torque. The proposed AMT control strategies found in the literature [14, 15, 13] use both controls possibilities. Although this approach is quite attractive, the engine torque is subject to many hard constraints and has much slower dynamics compared to the clutch's, particularly in the case of gasoline engines. The proposed strategy, therefore, aims to assure a comfortable engagement using as only control variable the clutch torque considering the engine torque as a known non-controllable input of the system whose evolution can be described using a homogeneous linear system.

In the following sections a simple output feedback PI controller based on the clutch sliding speed is proposed. This control scheme shows a strong sensitivity to the changes of the friction coefficient and needs the introduction of an additional feedback loop on the engine torque to compensate for the unwanted reduction of the transmission damping coefficient. The most important finding, however, is that even if the PI controller satisfies the GV no-lurch condition some residual oscillations are present even in the nominal case. This fact has motivated a deeper analysis of the synchronization resulting in the extension of the necessary no-lurch condition to a sufficient and necessary *ideal synchronization* condition. The last part of this chapter is devoted to the description of a finite-time optimal control scheme that satisfies this more general synchronization condition.

4.2 Synchronization Assistance Assuring the *GV No-lurch Condition*

4.2.1 Control Law

According to the Glielmo and Vasca's *GV no-lurch condition* the level of the driveline oscillations is controlled by the amplitude of the clutch sliding speed derivative $\dot\omega'_e - \dot\omega'_g$ at the synchronization.

Defining $y_1 = \omega'_e - \omega'_g$ we have from Equations 2.2a, 2.2b, 2.2c, 2.2d, and 2.6

$$\dot{y}_1 = -\frac{\gamma}{J_{t1}} F_n + \delta(t) \tag{4.1}$$

$$J'_g \dot{\omega}'_g = \gamma F_n - k_t \theta' - \beta'_t(\omega'_g - \omega'_v) \tag{4.2}$$

$$J'_v \dot{\omega}'_v = k'_t \theta' + \beta'_t(\omega'_g - \omega'_v) \tag{4.3}$$

$$\dot{\theta}' = \omega'_g - \omega'_v, \tag{4.4}$$

where:

$$J_{t1} = \frac{J'_e J'_g}{J'_e + J'_g} \qquad \delta(t) = \frac{k'_t}{J'_g} \theta' + \frac{\beta'_t}{J'_g} (\omega'_g - \omega'_v) + \frac{1}{J'_e} \Gamma_e.$$

The simple output feedback law

$$F_n = \frac{J_{t1}}{\gamma} (k_p y_1 + \delta(t)),$$

gives a sliding speed dynamic $\dot{y}_1 = -k_p y_1$ converging exponentially to zero thus satisfying the GV no-lurch condition.

This control law reduces sensibly the amplitude of the driveline oscillations but, even in the nominal case, some residual oscillations are present as it can be seen in Figure 4.3. Outside nominal conditions the control law shows a strong sensitivity to an error in the friction coefficient, particularly an overestimation of this parameter can lead to a failed synchronization (Figure 4.4).

The control law

$$F_n = \frac{J_{t1}}{\gamma} \left(k_p y_1 + k_i \int y_1 dt + \delta(t) \right)$$

introduces an integral component that increases the control robustness. The GV no-lurch condition is no longer satisfied; the residual oscillations have an amplitude proportional to the integral coefficient k_i. A value of k_i sufficiently high to assure a correct robustness of the controller induces a level of residual oscillations similar to the one measured in the absence of any synchronization-assistance strategy.

4.2.2 Feedback Effects and Engine Torque Control

The transfer function of the transmission torsion $\theta(s)$ for the previous system is

$$\theta'(s) = \frac{\frac{\gamma}{J'_g} F_n(s)}{s^2 + \frac{\beta'_t}{J_{t2}} s + \frac{k'_t}{J_{t2}}}. \tag{4.5}$$

This equation highlights the driveline damped oscillation mode.

The proportional feedback law $F_n = \frac{J_{t1}}{\gamma} (k_p y_1 + \delta(t))$ changes the transfer function to

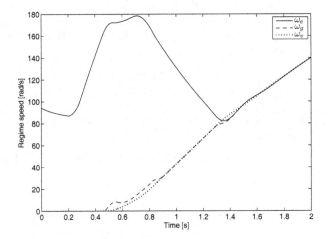

Figure 4.3. Standing-start simulation on flat ground with a synchronization assistance strategy based on a simple proportional feedback law

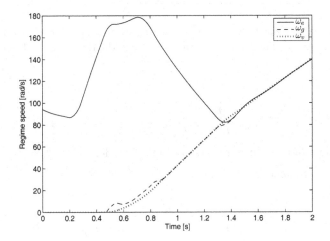

Figure 4.4. Standing-start simulation on flat ground with a synchronization-assistance strategy based on a simple proportional feedback law. A 5% error in the friction coefficient γ causes a failed synchronization.

$$\theta'(s) = \frac{\frac{J_{t1}}{J'_g}\left(k_p y_1 + \frac{\Gamma_e}{J'_e}\right)}{s^2 + \beta'_t \left(\frac{1}{J_{t2}} - \frac{J_{t1}}{J''^2_g}\right)s + k_t \left(\frac{1}{J_{t2}} - \frac{J_{t1}}{J''^2_g}\right)},$$

(4.6)

where

$$J_{t2} = \frac{J'_e J'_g}{J'_e + J'_g}.$$

The damping coefficient of Equation 4.6 has decreased due to the feedback action on the sliding speed.

An additional feedback loop on the engine torque Γ_e, as shown in Figure 4.5, with a control law

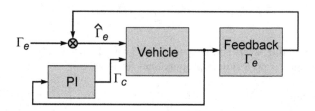

Figure 4.5. Structure of the PI control with a sliding speed feedback and pole placement through engine torque control to increase the damping coefficient

$$\hat{\Gamma}_e = \Gamma_e - \frac{J'_e J'_g}{J_{t1}}\varepsilon\left(\omega'_g - \omega'_v\right) = \Gamma_e - \frac{J'_e J'_g}{J_{t1}}\varepsilon\theta s$$

allows modification of the driveline oscillation mode.

The resulting transfer function is:

$$\theta'(s) = \frac{\frac{J_{t1}}{J_g}k_p y_1}{s^2 + \left[\beta_t \left(\frac{1}{J_{t2}} - \frac{J_{t1}}{J^2_g}\right) + \varepsilon\right]s + k_t \left(\frac{1}{J_{t2}} - \frac{J_{t1}}{J^2_g}\right)}.$$

The value of ε that gives a damping coefficient $\zeta = 1$ is

$$\varepsilon = 2\sqrt{k_t \left(\frac{1}{J_{t2}} - \frac{J_{t1}}{J^2_g}\right)} - \beta_t \left(\frac{1}{J_{t2}} - \frac{J_{t1}}{J^2_g}\right).$$

This strategy is very effective in simulation, as can be seen in Figure 4.6, but is dependent on a feedback on the engine torque. As previously mentioned, the engine torque control is subject to hard technological constraints and has to be a compromise with other vehicle performances, like fuel efficiency and emission control. All these constraints, not included in the previous simulation due to their complexity, could lead to quite a lower performance of this strategy. The active damping strategy included in all mass-produced vehicles, in fact, is quite similar in conception but has a limited impact due to the aforementioned limitations and the unavailability of the gearbox primary shaft speed ω'_g.

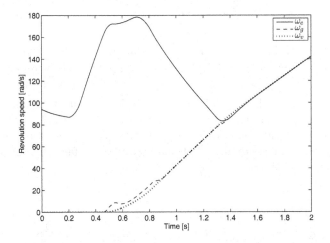

Figure 4.6. Standing-start simulation on flat ground with a synchronization-assistance strategy based on a PI controller with an additional feedback loop on the engine torque

4.3 GV No-lurch Condition Limitations

The analysis of the proposed synchronization assistance based on a simple proportional feedback has shown that respect of the GV no-lurch condition, *i.e.* a zero time derivative of the clutch sliding speed at synchronization, does not assure a lurch-free engagement.

Defining $y_1 = \omega'_e - \omega'_g$ and $y_2 = \omega'_g - \omega'_v$ we have from Equations 2.2a, 2.2b, 2.2c, and 2.2d

$$\dot{y}_1 = \frac{1}{J'_e}\Gamma_e - \frac{1}{J'_{t1}}\Gamma_c + \frac{1}{J'_g}\left(k'_t\theta' + \beta'_t y_2\right) \tag{4.7}$$

$$\dot{y}_2 = \frac{1}{J'_g}\Gamma_c - \frac{1}{J_{t2}}\left(k'_t\theta' + \beta'_t y_2\right) \tag{4.8}$$

$$\dot{\theta}' = y_2. \tag{4.9}$$

The dynamic system defined by the previous equations has a set of equilibrium points depending on the value of the engine torque Γ_e

$$y_{2\,eq} = 0 \tag{4.10}$$

$$\Gamma_{c\,eq} = \frac{J'_g + J'_v}{J'_e + J'_g + J'_v}\Gamma_e \tag{4.11}$$

$$\theta'_{eq} = \frac{1}{k'_t}\frac{J'_v}{J'_g + J'_v}\Gamma_c. \tag{4.12}$$

Ideally, at the synchronization instant t_s the three masses composing the simplified model have the same speed $w'_e = w'_g = w'_v = w$ and the same acceleration $\dot{w}'_e = \dot{w}'_g = \dot{w}'_v = \dot{w} = \frac{\Gamma_e}{J'_e + J'_g + J'_v}$, i.e. the system defined by Equations 4.7, 4.8, and 4.9 reaches its equilibrium point at t_s. This implies that $y_1(t_s^+) = w'_e(t_s^+) - w'_g(t_s^+) = 0$ but also that $y_2(t_s^+) = w'_g(t_s^+) - w'_v(t_s^+) = 0$. The GV no-lurch condition does not impose this second condition, as can be seen in Figure 4.7; the residual oscillations are due to the fact that the driveline did not reach its equilibrium point on t_s.

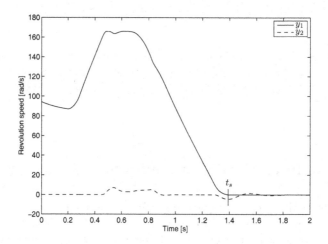

Figure 4.7. Plot of the y_1 and y_2 speed differences for a standing-start simulation with a synchronization-assistance strategy based on a simple proportional feedback. Although the GV no-lurch condition is met at t_s, $y_2(t_s) \neq 0$ causes some residual oscillations.

In energy terms, $y_2 = 0$ is equivalent to avoiding excessive potential elastic energy storage in the transmission stiffness. This energy is defined as $E_c = 1/2k_t\theta^2$ and its time derivative is $\dot{E}_c = k_t\theta\dot{\theta} = k_t\theta y_2$. In an equilibrium situation we have $\dot{E}_c = 0$ and

$$\theta = \frac{1}{k_t}\frac{J_v\Gamma_e}{J_e + J_g + J_v},$$

which gives $y_2 = 0$.

4.4 Synchronization Assistance with Ideal Engagement Conditions

4.4.1 Principle

The synchronization assistance has to reach the equilibrium point defined by

$$y_1(t_s) = 0 \qquad\qquad (4.13)$$

and Equations 4.10, 4.12, and 4.11. In order to avoid an excessive wear of the friction pads and to simplify the safety supervision of the strategy it is interesting to have a system assuring the end of the engagement in a pre-defined time.

Since the trajectory reaching the final equilibrium point has to simultaneously assure a good level of comfort, avoid excessive strain on the hydraulic actuator and limit the dissipated energy in the clutch, a finite-time optimal control strategy is the most adapted solution.

4.4.2 Cost Function

The quadratic cost function, which will be minimized by the optimal control, is defined using as reference the simplified driveline model given by Equations 4.7 - 4.9.

Under the simplifying hypothesis of a positive clutch sliding speed y_1, the relation between the normal force F_n exerted on the friction surfaces and the transmitted clutch torque is simply $\Gamma_c = \gamma F_n$. This hypothesis, which allows to the clutch to be considered as a torque actuator limits the validity domain of the control strategy; the respect of the $y_1 > 0$ condition can be either verified by inspection after the solution of the optimal control problem or embedded in the control problem itself by an additional inequality constraint.

To minimize the dissipated energy and avoid a high jerk the squared values of y_1 and y_2 are weighted. The first speed difference is proportional to the dissipated energy, while the second is proportional to the vehicle jerk.

The control action has to be weighted to form a well-posed optimal control problem. Given the physical structure of the clutch it is much more interesting to weight, instead of the transmitted clutch torque Γ_c, its time derivative since this quantity is related, through the non-linear washer-spring characteristic, to the slew rate of the hydraulic actuator.

In order to introduce the time derivative of the clutch torque an additional state is added to the simplified system.

$$\dot{\Gamma}_c = u. \tag{4.14}$$

The complete system in matrix form is

$$\dot{x} = \mathbf{A}x + \mathbf{B}_e\Gamma_e + \mathbf{B}_c u, \tag{4.15}$$

where

$$x = \begin{bmatrix} y_1 & y_2 & \theta & \Gamma_c \end{bmatrix}^T$$

$$\mathbf{A} = \begin{bmatrix} 0 & \frac{\beta'_t}{J'_g} & \frac{k'_t}{J'_g} & -\frac{1}{J_{t1}} \\ 0 & -\frac{\beta'_t}{J_{t2}} & \frac{k'_t}{J_{t2}} & \frac{1}{J'_g} \\ 0 & 1 & 0 & 0 \\ 0 & 0 & 0 & 0 \end{bmatrix} \quad \mathbf{B}_e = \begin{bmatrix} \frac{1}{J'_e} \\ 0 \\ 0 \\ 0 \end{bmatrix} \quad \mathbf{B}_c = \begin{bmatrix} 0 \\ 0 \\ 0 \\ 1 \end{bmatrix}.$$

The weighting of y_1, y_2 and u give the following quadratic cost function

$$J[y_1, y_2, u] = \frac{1}{2}\int_{t_0}^{t_s} \left[y_1^2(t) + a\, y_2^2(t) + b\, u^2(t) \right] dt$$

or, in matrix form,

$$J[x, u] = \frac{1}{2}\int_{t_0}^{t_s} \left[x^T\mathbf{Q}x + u^T\mathbf{R}u \right] dt, \tag{4.16}$$

where

$$\mathbf{Q} = \begin{bmatrix} 1 & 0 & 0 & 0 \\ 0 & a & 0 & 0 \\ 0 & 0 & 0 & 0 \\ 0 & 0 & 0 & 0 \end{bmatrix} \quad \mathbf{R} = [b].$$

4.4.3 Optimal Problem Formulation

Having defined the system dynamic equations and a quadratic cost function the synchronization-assistance strategy can be formalized as a finite-time optimal control problem.

Find $u(t)$ over the interval $T = [t_0, t_s]$ that minimizes the quadratic cost function (4.16) such that the system dynamic equations (4.15), the initial conditions $x(t_0) = x_0$ and the final conditions $x(t_s) = x_s$, defined as a function of the engine torque Γ_e by Equations 4.13, 4.10, 4.12, and 4.11, are satisfied.

4.4.4 Linear Quadratic Optimal Control

The linear quadratic (LQ) approach is the standard solution of the optimal control problem of a linear system with a quadratic cost function over both infinite and finite time intervals.

Considering the finite-time case, this method defines u over the time interval T minimizing the quadratic cost function

$$J(u) = \int_T \left(y^T \mathbf{Q} y + u^T \mathbf{R} u \right) \mathrm{d}t + y(t_s)^T \mathbf{F} y(t_s),$$

under the constrain of the dynamic equation

$$\dot{x} = \mathbf{A}x + \mathbf{B}u$$
$$y = \mathbf{C}x \qquad .$$

In this formulation the final state cannot be prescribed but can be forced through a heavy weighting of the final state.

The previous matrix formulation (4.15) cannot be directly used since the engine torque Γ_e is, by design choice, a known exogenous input of the system. The case of a controllable engine torque has already been analyzed in the literature [16].

Under the hypothesis of a constant engine torque Γ_{e0} on the optimization interval T, a change of variable can be used to write the dynamics of the system around the equilibrium point

$$x_{eq} = \left[0 \; 0 \; \frac{1}{k_t'} \frac{J_v'}{J_e' + J_g' + J_v'} \Gamma_{e0} \; \frac{J_g' + J_v'}{J_e' + J_g' + J_v'} \Gamma_{e0} \right]^T .$$

The combination of the cost function (4.16) and the final state conditions gives

$$\mathbf{Q} = \mathrm{diag}\left(\begin{bmatrix} 1 & a & 0 \end{bmatrix} \right) \quad \mathbf{R} = \begin{bmatrix} c \end{bmatrix} \quad \mathbf{F} = \frac{1}{\epsilon} \mathrm{diag}\left(\begin{bmatrix} 1 & 1 & 1 \end{bmatrix} \right),$$

where ϵ is small enough to assure a final state close to the ideal synchronization condition.

The resulting trajectories, shown in Figures 4.8 and 4.9, have a final state close to the ideal synchronization conditions but do not satisfy exactly the final condition in spite of the extremely heavy final state weighting.

4.4.5 Optimal Control by Differential Analysis

Dynamic Lagrangian Multipliers

In order to impose precisely a prescribed final state we must resort to the more general differential analysis theory [1].

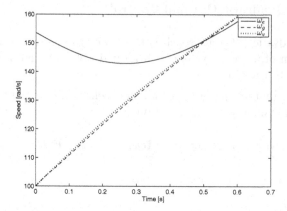

Figure 4.8. Optimal synchronization trajectories obtained through an LQ formulation with $a = 1$, $b = 10^{-2}$ and $\epsilon = 10^{10}$

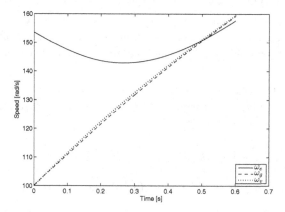

Figure 4.9. Optimal synchronization trajectories obtained through an LQ formulation with $a = 1$, $b = 10^{-2}$ and $\epsilon = 10^{10}$

The dynamic Lagrangian multipliers method is an extension to the constrained dynamic optimization problem of the standard Lagrangian multipliers solution of constrained static optimization problems. The working principle is the same: the constrained optimization of the free variable x is transformed, through the introduction of the λ Lagrangian multipliers, in an equivalent *dual problem* consisting of an unconstrained optimization of the free variables x and λ. It can be shown that the solution to the dual problem is, in the general case, an upper bound of the original problem solution; in particular if the primary problem is convex, as in our case, its solution is coincident with that of the dual problem; this property is called *strong duality*.

The solution of the dual problem and, since the strong duality conditions are met, of the primary problem is expressed as the solution of a two-point boundary-value problem (TPBVP).

In particular the input trajectory solving the optimal control problem

Find $u(t)$ over the interval $T = [t_0, t_s]$ that minimizes the quadratic cost function (4.16) such that the system dynamic equations (4.15), the initial conditions $x(t_0) = x_0$ and the final conditions $x(t_s) = x_s$, defined as a function of the engine torque Γ_e by Equations 4.13, 4.10, 4.12, and 4.11, are satisfied.

is defined by the following TPBVP

$$\dot{x} = \mathbf{A}x + \mathbf{B}_e\Gamma_e + \mathbf{B}_c u \tag{4.17a}$$

$$\dot{\lambda} = -\mathbf{Q}x - \mathbf{A}^T\lambda \tag{4.17b}$$

$$u = -\mathbf{R}^{-1}\mathbf{B}_c^T\lambda. \tag{4.17c}$$

For the mathematical justification of this result, from the problem definition to the derivation of the TPBVP, we invite the reader to consult Appendix A.1.

Two-point Boundary-value Problem by Shooting Method

The optimal control trajectory is defined as the solution of a TPBVP. The peculiarity of this problem is that the boundary-value constraints insuring the uniqueness of the solution of the differential equation system are not the complete initial state vector, as in the usual case, but half of the initial state vector and half of the final state vector.

The standard numerical solution for this kind of problem is a numeric iterative method called *shooting*. Defining the operator

$$F(\lambda_0) \rightarrow x(t_s)$$

transforming the initial value of the co-states

$$\lambda_0 = \lambda(t_0) = \begin{bmatrix} \lambda_1(t_0) & \lambda_2(t_0) & \lambda_3(t_0) & \lambda_4(t_0) \end{bmatrix}^T$$

into the state final value

$$x(t_s) = \begin{bmatrix} y_1(t_s, \lambda_0) & y_2(t_s, \lambda_0) & \theta(t_s, \lambda_0) & \Gamma_c(t_s, \lambda_0) \end{bmatrix}^T,$$

the shooting method is simply finding the root $\tilde{\lambda}_0$ to the following vectorial equation

$$F(\lambda_0) - x_s = 0, \tag{4.18}$$

i.e. finding the initial co-state values giving a system evolution reaching the desired final states.

The knowledge of the initial values of the co-state transforms the TPBVP in a simple initial-value problem (IVP) easily solved by numerical forward integration.

Since the search for the roots of Equation 4.18 is done by iteration and each evaluation of the $F(\lambda_0)$ operator implies a complete forward integration of the differential system over the $T = [t_0, t_s]$ time interval, this method is extremely computing-power intensive. Its performances and sometimes its convergence are highly dependent on the initial guess of λ with which the iteration is started. Some variants, like the *multiple shooting* method, ease some of these difficulties but this approach remains a not very elegant hit-and-miss solution compared to the more refined ones outlined in the following sections.

Two-point Boundary-value Problem by Matrix Exponential

Substituting Equation 4.17c in Equation 4.17a the TPBVP can be written as

$$\frac{\mathrm{d}}{\mathrm{d}t}\begin{bmatrix} x \\ \lambda \end{bmatrix} = \begin{bmatrix} \mathbf{A} & -\mathbf{B}_c\mathbf{R}^{-1}\mathbf{B}_c^T \\ -\mathbf{Q} & -\mathbf{A}^T \end{bmatrix}\begin{bmatrix} x \\ \lambda \end{bmatrix} + \begin{bmatrix} \mathbf{B}_e \\ 0 \end{bmatrix}\Gamma_e$$

or, in a more compact form,

$$\dot{z} = \mathbf{A}_z z + \mathbf{B}_z \Gamma_e. \tag{4.19}$$

Under the hypothesis of a constant engine torque Γ_e over the optimal control activation interval,[1] this non-controlled input can be considered as a constant additional state as shown for the LQ method. The TPBVP equation (4.19) becomes a simple homogeneous differential equation

$$\dot{w} = \mathbf{A}_w w,$$

where

$$w = \begin{bmatrix} x & \lambda & \Gamma_c & \Gamma_e \end{bmatrix}$$

$$\mathbf{A}_w = \begin{bmatrix} \mathbf{A}_z & \mathbf{B}_z \\ 0 & 0 \end{bmatrix},$$

with an additional initial condition $\Gamma_e(t_0) = \Gamma_{e0}$.

Since the system is linear

$$w(t_s) = e^{\mathbf{A}_w t_s} w(t_0) = \Phi_{t_s} w(t_0),$$

---

[1] Experimental evidence validates this hypothesis for an activation interval of about 0.5 s.

or, in a more compact form,

$$
\begin{bmatrix} x(t_s) \\ \lambda(t_s) \\ \Gamma_e(t_s) \end{bmatrix} = \begin{bmatrix} \varphi_{11} & \varphi_{12} & \varphi_{13} \\ \varphi_{21} & \varphi_{22} & \varphi_{23} \\ \varphi_{31} & \varphi_{32} & \varphi_{33} \end{bmatrix} \begin{bmatrix} x(t_0) \\ \lambda(t_0) \\ \Gamma_e(t_0) \end{bmatrix}, \tag{4.20}
$$

where $\Gamma_e(t_s) = \Gamma_e(t_0)$ by hypothesis. This line of reasoning actually holds for any evolution of the engine torque that can be described as a homogeneous linear system. For simplicity, we will limit our analysis to the simplest case of a constant value.

On imposing the boundary conditions the first line defines the following linear system

$$
\varphi_{12}\lambda_0 = x_s - \varphi_{11}x_0 - \varphi_{13}\Gamma_{e0}, \tag{4.21}
$$

which defines the initial co-state values λ_0 as a linear combination of the initial and final state values x_0, x_s and the constant engine torque Γ_e.

The matrix exponential does not have a closed-form expression but for the particular case of a diagonal matrix

$$
\mathbf{D} = \mathrm{diag}[d_1, d_2, ...d_n],
$$

where

$$
e^{\mathbf{D}} = \mathrm{diag}[e^{d_1}, e^{d_2}, ...e^{d_n}].
$$

The matrix $\mathbf{A}_w$ has independent eigenvalues and, therefore, is diagonalizable; *i.e.* that exists a matrix $\mathbf{V}$ composed by $\mathbf{A}_w$ eigenvectors arranged in columns such that $\mathbf{D}_w = \mathbf{V}^{-1}\mathbf{A}_w\mathbf{V}$ is diagonal. Since

$$
e^{\mathbf{A}_w t} = \mathbf{V}e^{\mathbf{D}_w t}\mathbf{V}^{-1}
$$

we might be tempted to use this relation to obtain a closed-form solution of the linear system. Unfortunately, this is not feasible because it simply shifts the problem to the determination of the closed form of the $\mathbf{V}$ matrix that does not exist in the general case for a matrix having a dimension bigger than three. The matrix exponential, therefore, must be approximated using one of the numerical algorithms of [26].

The linear system (4.21) has a solution if and only if the matrix φ_{12} is invertible, *i.e.* the TPBVP is well posed.

An initial-value problem

$$
\dot{x} = f(x, u)
$$
$$
x(t_0) = x_0
$$

is well posed if $f(x, u)$ is Lipschitz in x. A similar well-posedness simple condition, even limited to the linear case, does not exist for a general TPBVP. In our case, however, the TPBVP is an expression of the KKT optimality conditions.

The optimal control problem under analysis has a quadratic cost function with positively defined $\mathbf{Q}$ and $\mathbf{R}$ matrices over a convex domain since it is defined by a set of linear constraints. Under these conditions optimization theory guarantees the existence and uniqueness of the optimal solution. Moreover, since we have a strong duality and the Abadie constraint qualification is satisfied, every optimal solution must satisfy the KKT conditions and *vice versa*. The existence and uniqueness of the optimal solution therefore implies the existence and uniqueness of the TPBVP solution defined by the KKT conditions and, finally, the invertibility of the φ_{12} matrix.

Numerical Difficulties

Despite the fact that the invertibility of the φ_{12} matrix is theoretically guaranteed, numerical difficulties can be experienced while trying to solve the linear system (4.21) for time intervals longer than one second in combination with particularly stiff transmission shafts and low gearbox inertias.

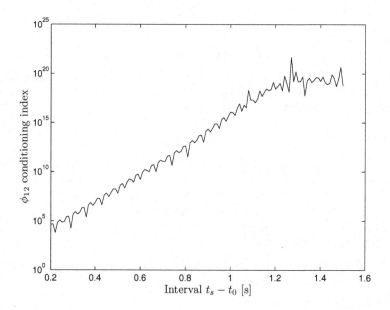

Figure 4.10. Conditioning index of the φ_{12} matrix as a function of the length of the optimal control activation interval

Linear geometry theory states that a square matrix A is either invertible or not with a clear cut distinction between the two cases. When using a finite precision arithmetic this distinction is more blurred. The conditioning index

$$cond_p(A) = \|A^{-1}\|_p \|A\|_p$$

is a measure of a matrix invertibility. The conditioning index for the quadratic norm can also be defined through the singular values $\sigma(A)$

$$cond_2(A) = \frac{\sigma_{max}(A)}{\sigma_{min}(A)}$$

or, for a normal matrix, through its eigenvalues $\lambda(A)$

$$cond_2(A) = \frac{\lambda_{max}(A)}{\lambda_{min}(A)}.$$

This index has a value between 1 and infinity. The higher this value is the more difficult is the matrix inversion using finite precision arithmetic.

Considering the solution of a linear system $\mathbf{A}x = b$, $m = \log_{10}(cond_2(A))$ is roughly the number of significative digits lost in the solution. Since standard numerical packages like MATLAB® have an internal double precision float representation of real numbers, $i.e.$ roughly 15 significative digits, it can be seen in Figure 4.10 that the φ_{12} matrix is not invertible for activation intervals longer than one second.

This limitation does not affect the synchronization-assistance strategy that has short activation intervals but makes it impossible to obtain a complete optimal standing-start. A deeper analysis of the algorithms used for the calculation of the matrix exponential itself [25, 26] are of little avail since this difficulty is intrinsic to the finite-time optimal control problem. An alternative solution of the TPBVP based on the generating functions has been taken into consideration and tested but its complexity makes this solution of little practical interest. For the exact details of this approach the reader is invited to see Appendix A.2.

More practical solutions are increasing the significative digits in the real numbers representation using an arbitrary precision arithmetic library or solving the optimization problem through reconduction to a standard quadratic programming formulation. While the first solution simply allows for longer activation intervals by using an extended precision the second solution actually sidesteps the problem transforming the dynamic optimization problem in a static optimization problem. Both solutions still have practical limitations on the length of the optimization horizon due to memory and run-time constraints but these are of no concern in our case. In the case of an extremely long activation time the only available solution is the complex generating function approach previously discussed.

Activation Interval Length and Triggering Threshold

The dynamic Lagrangian approach gives a solution to the optimal control problem for a given initial condition x_0 and activation interval $t_s - t_0$. The prescribed final condition $x_s = f(\Gamma_e(t_s))$ is obtained from the initial condition vector under the hypothesis of a constant engine torque. The choice of the clutch slipping speed threshold $y_1(t_0)$ triggering the activation of the synchronization-assistance strategy and the activation interval length $t_s - t_0$ has yet to be investigated.

The optimal control problem, in the formulation specified in the previous sections, has a solution for every positive value of $y_1(t_0)$ and $t_s - t_0$ but an excessively long activation interval compared to the triggering slipping speed lengthens the engagement without providing any particular advantage thus reducing the comfort. On the other hand, too short an activation interval induces an increase of the clutch torque before its reduction, causing a highly uncomfortable vehicle oscillation. In order to better understand the relation between these two quantities the engagement of a torsion-free driveline is considered.

Defining $J_1 = J'_g + J'_v$, the equations of a two degrees of freedom torsion free-driveline are

$$J'_e \dot{\omega}'_e = \Gamma_e - \Gamma_c$$
$$J_1 \dot{\omega}'_v = \Gamma_c,$$

which give for the sliding speed the following equation

$$\dot{y}_1 = \frac{1}{J'_e}\Gamma_e - \left(\frac{1}{J'_e} + \frac{1}{J_1}\right)\Gamma_c. \tag{4.22}$$

Given the initial condition $y_1(t_0)$, $\Gamma_e(t_0)$ and $\Gamma_c(t_0)$, if no intervention is made, i.e. $\Gamma_c(t) = \Gamma_c(t_0)$, the synchronization will happen after a time interval

$$\Delta t = \frac{y_1(t_0)}{\frac{1}{J'_e}\Gamma_e(t_0) - \left(\frac{1}{J'_e} + \frac{1}{J_1}\right)\Gamma_c(t_0)}.$$

If we consider only $\Gamma_c(t) \le \Gamma_c(t_0)$ for $t \in [t_0, t_s]$ the following relation between $y_1(t_0)$ and $t_s - t_0$ holds

$$y_1(t_0) = \alpha(t_s - t_0)\left(\frac{1}{J'_e}\Gamma_e(t_0) - \left(\frac{1}{J'_e} + \frac{1}{J_1}\right)\Gamma_c(t_0)\right), \tag{4.23}$$

with $\alpha \in (0, 1)$. In Figure 4.11 the optimal trajectories for several values of α ranging from 0.2 to 0.8 are shown for an activation interval of $t_s - t_0 = 0.5$ seconds, an initial engine torque $\Gamma_e(t_0) = 50$ Nm and an initial clutch torque $\Gamma_c(t_0) = 60$ Nm on a Clio II driveline. We note that even if there exists by construction a feasible solution respecting the constraint $\Gamma_c(t) \le \Gamma_c(t_0)$, the optimal solution might differ from it for high values of the α parameter. On the other hand, a very low α value can induce an excessive reopening of the clutch. An α value of 0.5 avoids these two extremes and minimizes the actuator activity.

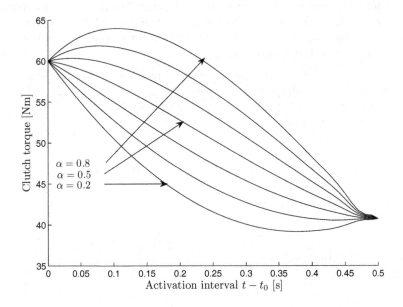

Figure 4.11. Clutch torque optimal engagement trajectories for a Clio II over an activation interval of $t_s - t_0 = 0.5$ s, $\Gamma_e(t_0) = 50$ Nm, $\Gamma_c(t_0) = 60$ Nm and $\alpha \in [0.2, 0.8]$

Model Validity Domain

Even if all the optimal trajectories calculated are within the simplified model validity domain, *i.e.* a positive clutch sliding speed y_1, this condition is not assured by any mathematical property.

To guarantee the physical feasibility of the optimal trajectories either they are obtained offline and validated by inspection before programming them on the vehicle, or an additional inequality constraint is added to the optimal problem formulation.

Adding to the previous optimal control problem the feasibility inequality constraint $y_1 \geq 0$ and the comfort-induced constraint $u \leq 0$ motivated by the results presented in the previous section, we have

Find $u(t)$ over the interval $T = [t_0, t_s]$ that minimizes the quadratic cost function (4.16) such that the system dynamic equations (4.15), the initial conditions $x(t_0) = x_0$, the final conditions $x(t_s) = x_s$, defined as a function of the engine torque Γ_e by Equations 4.13, 4.10, 4.12, and 4.11, and the inequality conditions $y_1 \geq 0$ and $u \leq 0$ are satisfied.

The dynamic Lagrangian multipliers method cannot be applied to this optimal control problem due to the non-constructive nature of the KKT conditions induced by the inequality constraints. The reader is invited to refer to Appendix A.1.2 for a more detailed explication.

An alternative solution to the optimal control problem is thus needed. In the following section we will see how the quadratic problem formulation allows us to both introduce inequality constraints and avoid the numerical difficulties previously highlighted.

4.4.6 Optimal Control by Quadratic Programming

The static optimization problem consists in finding the minimizing vector v with respect to

$$J[v] = v^T \mathbf{H} v + v^T f,$$

under the following constraints

$$A_{eq} v = b_{eq},$$

and

$$A_{in} v = b_{in}$$

is called the quadratic programming problem. Since many real-life situations involve an optimization problem that can be stated using a quadratic programming formulation its solution methods are well known and easily available in most optimization libraries.

The dynamic optimization problem

> Find $u(t)$ over the interval $T = [t_0, t_s]$ that minimizes the quadratic cost function (4.16) such that the system dynamic equations (4.15), the initial conditions $x(t_0) = x_0$, the final conditions $x(t_s) = x_s$, defined as a function of the engine torque Γ_e by Equations 4.13, 4.10, 4.12, and 4.11, and the inequality conditions $y_1 \geq 0$ and $u \leq 0$ are satisfied.

can be expressed using a QP formulation; introducing a sampling finding the optimal continuous $u(t)$ function can be thought of as optimizing the vector $\bar{u}$ formed by the samples u_k at sampling instants t_k.

The interested reader can find the exact details concerning this transformation together with the definition of the matrices and vectors $\mathbf{H}$, f, A_{eq}, b_{eq}, A_{in} and b_{in} in Appendix A.3.

The resulting optimal trajectories from the solution of the QP problem for the same conditions as in Figure 4.11 are shown in Figure 4.12. The effect of the $u \leq 0$ inequality constraint is very clear for the trajectories calculated for the extreme values of the α parameter. The clutch torque in these cases is either saturated to the initial value in the first part of the trajectory or to the final value for the last part. The other inequality constraint, on the other hand, being inactive since it was respected even in the unconstrained case, has no influence whatsoever on the resulting trajectories.

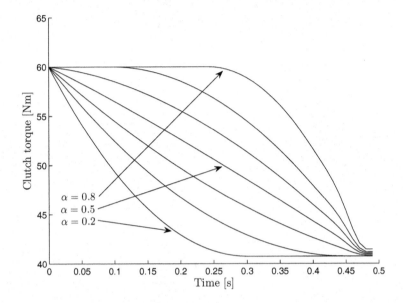

Figure 4.12. Clutch torque optimal engagement trajectories for a Clio II over an activation interval of $t_s - t_0 = 0.5s$, $\Gamma_e(t_0) = 50$ Nm, $\Gamma_c(t_0) = 60$ Nm and $\alpha \in [0.2, 0.8]$ and subject to the inequality contraints $u \leq 0$ and $y_1 \geq 0$

5

Optimal Standing-start

5.1 Principle

The solution of the optimal control problem through the use of a quadratic programming formulation has raised the activation time limitation imposed by the solution of the TPBVP over long intervals, allowing calculation of a complete optimal standing-start trajectory. This kind of solution could be used for the control of a AMT or a clutch-by-wire system where the clutch has to be controlled by the gearbox control unit from the very beginning of the engagement.

In the previous chapter the solution of the optimal control problem has been obtained under the hypothesis of a constant engine torque. This condition, perfectly reasonable for a short activation time over the last part of the engagement, is very hard if not downright impossible to assure when considering a whole standing-start operation. The possibility of having an engine torque evolution described by an homogeneous linear system could partially solve this difficulty but does not allow for a change of the driver's wish during the engagement. Taking into account these changes not only increases the driving comfort but is essential for security since the driver has to be always able to intervene in the vehicle behavior in order to react to a rapid change in his environment.

To meet the opposing needs of taking into account the driver's input and having some simple hypothesis on future input allowing for an optimal planning, the trajectory is not computed once and for all at the beginning of the engagement and then simply tracked by feedback, but periodically updated to follow the changes in the driver's input and the actual behavior of the vehicle.

5.2 Exact Dynamic Replanning

5.2.1 Model Predictive Control

The model-based predictive control (MPC) is a control strategy that, in its most general formulation, consists in solving an optimal control problem in QP formulation for a simplified and/or liberalized system over a finite time horizon of N_h samples and issuing the first N_c control samples before using the newly measured or estimated system state x_0 as the starting point for a new optimization.

Compared to the previous control schemes the MPC strategy shows two main advantages, the first being that a change in the driver's input can be readily taken into account in the next optimization. The second advantage is that trajectory stabilization does not need to be assured by an external feedback loop since each new optimal trajectory has as a starting point the measured or estimated system state x_0 that directly takes into account the actual system behavior.

5.2.2 Optimization Horizon Update

When updating the optimal control trajectory two choices concerning the control horizon are possible: either we keep a constant horizon of N_h samples or we reduce it by N_c samples to account for the time that has passed since the last optimization. The first option is called the *sliding horizon*, while the second is the *fixed horizon*; respectively, shown as A and B cases in Figure 5.1.

In a sliding-horizon MPC control strategy, by far the most common case, the following optimal QP control problem is formulated and solved for each update.

Find the vector
$$\bar{u} = \begin{bmatrix} u_0 \ \ldots \ u_{N_h-1} \end{bmatrix}^T$$

minimizing the quadratic cost function
$$J[\bar{u}] = \bar{u}^T \mathbf{H} \bar{u} + f \bar{u}$$

under the constraints
$$\mathbf{A}_{eq} \bar{u} = b_{eq}$$
$$\mathbf{A}_{in} \bar{u} = b_{in},$$

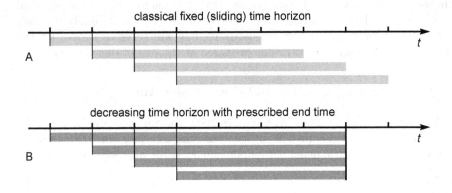

Figure 5.1. Optimization horizon update for a sliding-horizon (A) and a fixed-horizon (B) formulations

where $\mathbf{H}$, f, $\mathbf{A}_{eq}$, b_{eq}, $\mathbf{A}_{in}$ and b_{in} are defined according to Appendix A.3

The only element that changes between one optimization run and the other is the initial state vector x_0 that is part of the definition of f and b_{eq}. The optimization procedure can be thus seen as a mapping of the phase space of the initial state vectors x_0 onto the space $\mathbb{R}^{m \times N_c}$ of the first N_c optimal control samples $u_j \in \mathbb{R}^m$ with $j \in [1, N_c]$.

In the particular case of $N_c = 1$ it can be shown [17] that there exists a polyhedric partition of the phase space for which in every region $\mathcal{R}_k$ the following relation between the first optimal control sample u_0 and the initial state vector $x_0 \in \mathcal{R}_k$ holds:

$$u_0 = \mathbf{A}_k x_0 + b_k.$$

The problem of calculating this partition and the corresponding sets of matrices $\mathbf{A}_k$ and vectors b_k is called multi-parametric quadratic programming or the *mpQP* problem. This solution is particularly interesting for systems having few states and that can be controlled with a short prediction horizon otherwise the partition has so many regions that the computational cost of checking which region contains the initial state x_0 is higher than the one corresponding of a well-implemented QP algorithm.

Globally, the behavior of a sliding-horizon MPC strategy and of an infinite time optimal LQ controller, an exception is made for the constraints, are quite similar. In particular, due to the sliding horizon, no final state can be prescribed and no synchronization time is guaranteed. For this reason a fixed-horizon MPC strategy, allowing both for a specified final state and a fixed synchronization time, has been selected

In the following section a fixed-horizon MPC engagement control strategy will
be presented. Due to its computational cost this solution is clearly impossible to implement on the current hardware but will serve as a base for the
conception of a simplified strategy proposed at the end of this chapter.

5.2.3 *Model Predictive Control* Control Structure

The control structure of a MPC engagement control strategy for an AMT
vehicle is shown in Figure 5.2. In this case the interface with the driver is
limited to the throttle pedal whose position x_p is the only system input.

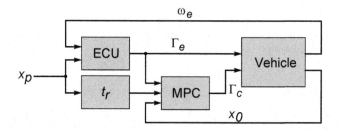

Figure 5.2. Control structure of an MPC engagement control strategy for a AMT
vehicle. The throttle pedal position x_p is the only system input, the engine control
unit *ECU* outputs the engine torque target Γ_e based on x_p and ω'_e. The control
horizon is a function of x_p and the clutch torque is controlled by the MPC strategy
based on the driveline state x_0, the control horizon length t_r and the engine torque.

The engine control unit (ECU) outputs an engine torque target Γ_e based on
the throttle pedal position x_p and engine speed ω'_e following the throttle look-
up table and the *ad hoc* strategies for comfort, fuel efficiency and emission
reduction.

The time-to-synchronization, *i.e.* the control horizon, t_r is given as

$$t_r = (1 - \alpha)t_f(x_p),$$

where α is the percentage of completed engagement, defined as

$$\alpha = \int_{t_0}^{t} \frac{1}{t_f(x_p(\tau))} d\tau,$$

and $t_f(x_p)$ is the total engagement time for a given throttle pedal position.
This relation, expressed by a simple look-up table, is the main tuning parameter for the vehicle brio.[1]

---

[1] The usual compromise between the vehicle brio and the lurch at synchronization
does not exist in this case since the MPC control is supposed to assure a lurch-free
synchronization in any case.

The MPC strategy obtains the clutch torque Γ_c, solving the optimal control problem

Find the function $u(\tau)$ defined over the time interval $T = [t, t + t_r]$ that minimizes

$$J[u] = \int_t^{t+t_r} \left[x^T \mathbf{Q} x + u^t \mathbf{R} u \right] \mathrm{d}t,$$

under the constraints

$$\dot{x} = \mathbf{A}x + \mathbf{B}u$$

$$x(t) = x_0 \quad x(t + t_r) = x_s(\Gamma_e(t))$$

$$u \leq 0 \quad y_1 \geq 0,$$

defined as a function of the time-to-synchronization t_r, the engine torque Γ_e supposed constant and the driveline state x_0. The solution of the optimal control problem is obtained by quadratic programming.

5.2.4 Results

The strategy had been tested only in simulation, first in a simplified case for testing the implementation of the strategy itself and the viability of a constant engine torque hypothesis updated every sample instant; then on a complete driveline model including a driver model and an engine torque-generation model as described in Section 2.2.1.

In the simplified case, shown in Figure 5.3, the engine torque is simply a saturated ramp and the engagement time $t_f = 2$ s is constant, which gives for the time-to-synchronization $t_r = 2 - t$.

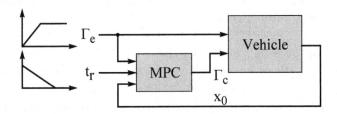

Figure 5.3. Structure of the simplified case used to validate the principle of the fixed-horizon MPC strategy

Figures 5.4 and 5.5 show the resulting trajectories for the simplified case. Initial conditions x_0 are marked by a star and the curves are the QP problem solutions obtained at each control step. The resulting complete trajectory, excepted for the final partial clutch reopening assuring a lurch-free engagement, is quite similar to the behavior of a human driver controlling the clutch pedal in an MT vehicle. The first few solutions have a synchronization engine speed that is too low and would lead to an engine stall. This is due to the fact that in this simplified case the synchronization time is kept constant to 2 s imposing quite a fast standing-start even for very low engine torques.

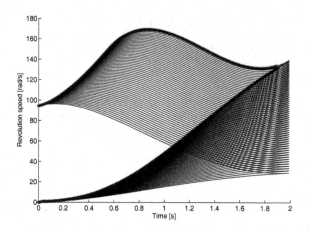

Figure 5.4. Standing-start simulation for a Clio II 1.5dCi (K9K) with a saturated ramp engine torque profile. The stars mark the optimization starting point.

Figures 5.6 and 5.7 show the results of a standing-start simulation with the MPC strategy on a complete driveline with a torque-generation model and a controlling driver model. The combined effect of the driver model's actions on the clutch pedal position x_p and the engine output torque variations induced by the torque-generation model induce strong variations on the engine torque profile. The MPC strategy copes correctly with these changes and assures a comfortable engagement respecting the ideal synchronization conditions.

The MPC strategy is therefore well adapted for this situation, but cannot be implemented directly on a vehicle due to the high computational cost involved in solving the QP problem each sample time.

The main controlling parameter for the execution time of a QP algorithm is the number of free variables subject to optimization, in our case given by the number of sample times before synchronization. Considering a 100 Hz controller and a slow, 3 s standing-start the first MPC cycles will have 300 free variables. A problem of this size has a solution time with the MATLAB® routine quadprog of more than 15 s on an *AMD Athlon64 3000+*.

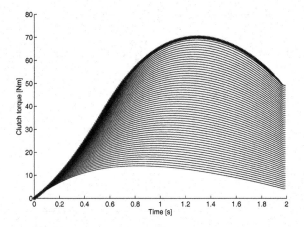

Figure 5.5. Standing-start simulation for a Clio II 1.5dCi (K9K) with a saturated ramp engine torque profile. The stars mark the optimization starting point for the optimal clutch torque trajectories drawn as a solid line.

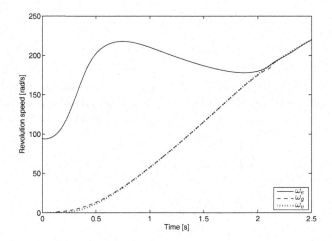

Figure 5.6. Standing-start simulation using the MPC strategy on a Clio II 1.5dCi (K9K) with a complete driveline with a torque-generation model and a controlling driver model. A comfortable engagement is assured in spite of the strong changes in the engine torque.

5.3 Simplified Dynamic Replanning

5.3.1 Segment-approximated Model Predictive Control

In the previous definition of the MPC strategy the control variable $u = \dot{\Gamma}_c$ was defined as the time derivative of the clutch torque to guarantee that the

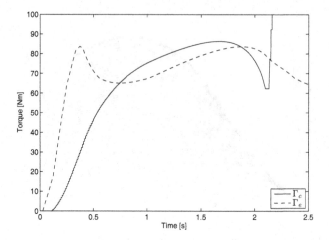

Figure 5.7. Standing-start simulation using the MPC strategy on a Clio II 1.5dCi (K9K) with a complete driveline with a torque-generation model and a controlling driver model. A comfortable engagement is assured in spite of the strong changes in the engine torque.

resulting Γ_c trajectory were continually differentiable, *i.e.* $\Gamma_c \in C^1$ and to weight the hydraulic actuator's slew rate [11, 12].

Thanks to the introduction of this additional derivative the zero holder sampling used to introduce a QP formulation can be viewed as limiting the search of the optimal solution to the class of curves composed by a finite number of segments:

$$\Gamma_c(t) = \begin{cases} \Gamma_{c0} + u_0 t & t \in T_1 \\ \Gamma_{c0} + u_0 \Delta t + u_1 t & t \in T_2 \\ \vdots \\ \Gamma_{c0} + \sum_{j=0}^{N-2} u_j \Delta t + u_{N-1} t & t \in T_N \end{cases}$$

$$T_j = [(j-1)\Delta t \; j\Delta t) \quad N = tr/\Delta t$$

where Δt is the sampling interval.

Seen from this angle, a heavy under-sampling of the previous optimal control problem constrains the optimal solution to be composed of a limited number N of segments. As can clearly be deduced from Figure 5.8 even very few segments can give a fairly accurate optimal trajectory.

In order to fully take advantage of this insight the MPC control frequency and the QP problem sampling frequency must be dissociated.

In the previous MPC formulation for every sampling interval Δt a new QP optimization problem with a prediction horizon of $N_h = tr/\Delta t$ is formulated and solved. The first control sample of the optimal solution is used and, after

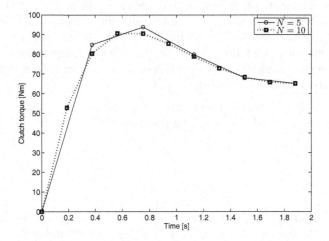

Figure 5.8. Example of two optimal trajectories composed by 5 and 10 segments

Δt s, this procedures is iterated, meaning that if the throttle pedal did not move, a new QP problem with $N_h - 1$ free variables is solved.

Obviously simply introducing a heavy under-sampling without any change in the control structure would greatly reduce the system performances. This difficulty can be solved by keeping the number of the QP problem free variables N constant independently of the optimization time horizon t_r and the MPC control sampling interval Δt. This scheme is equivalent to searching the optimal trajectory composed by a constant limited number N of segments independently of the effective engagement duration. Since the number of segment composing the trajectory is kept constant while the prediction horizon t_r gets shorter, the planned trajectory, initially quite coarse, gets increasingly more precise.[2]

5.3.2 State Vector Reduction

In the two previous formulation of the MPC strategy a complete knowledge of the driveline state x_0 has been implicitly admitted. On a standard AMT vehicle ω'_e, ω'_g and ω'_v speed are directly measured, Γ_e is estimated by the engine control unit and Γ_c is obtained by inversion of the clutch characteristic curve learned by the gearbox control unit. The missing element from the state vector x_0, *i.e.* the transmission shafts torsion θ', could be obtained through

---

[2] Actually, in the last N iterations of the strategy the segment approximated solution has a sampling interval shorter than that of the standard QP solution and thus is closer to the continuous optimal solution given by the dynamic Lagrangian multipliers method.

a simple driveline state observer. Since one of the main concerns while developing this MPC strategy has been to limit the computational cost, a further simplification of the driveline model has been studied. Ignoring the driveline torsion, on the other hand, does not further reduce the computational cost associated with the solution of the QP problem since the dynamic of the system has been embedded in the problem formulation itself.

Since the driveline torsion has been ignored the ideal synchronization conditions cannot be specified, thus limiting the solution to a sub-optimal engagement control.

5.3.3 Results

In order to compare the performances of the different control schemes three simulation runs representing a standing-start on flat ground with the same throttle pedal profile have been completed. The results are shown in Figures 5.9 and 5.10. While the segment approximation of the optimal trajectory has a very limited impact on the control performances despite its drastic reduction in computing cost, the state reduction is clearly sub-optimal and shows some residual oscillations after the engagement.

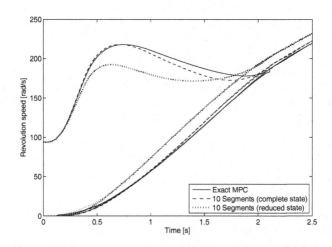

Figure 5.9. Engine, gearbox and vehicle equivalent speeds for a standing-start simulation of a Clio II on flat ground using the exact, segment-approximated and reduced-state segment-approximated MPC strategy

All these simulations show a good compensation of the changing engine torque due to use of the engine torque-generation model in spite of the constant-torque hypothesis introduced to define the QP problem.

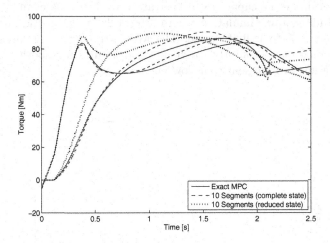

Figure 5.10. Engine and clutch torques for a standing-start simulation of a Clio II on flat ground using the exact, segment-approximated and reduced-state segment-approximated MPC strategy

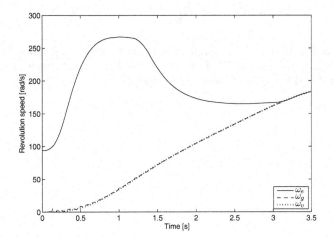

Figure 5.11. Engine, gearbox and equivalent vehicle speed simulation results for a standing-start simulation with a 10-segment MPC strategy; after 1.2 s of engagement the throttle position is reduced from 20% to half of its value

Finally, to evaluate the capacity of the strategy to react to a changing driver input a standing-start simulation where after 1.2 s the throttle position is reduced from 20% to half its value. The MPC strategy using the segment approximation reacts promptly to change and assures a smooth transition from quite a fast engagement to a slow one.

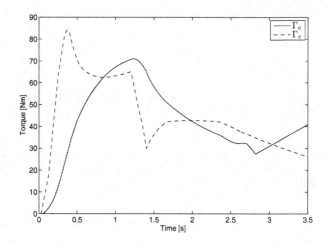

Figure 5.12. Engine and clutch torque simulation results for a standing-start simulation with a 10-segment MPC strategy; after 1.2 s of engagement the throttle position is reduced from 20% to half of its value

6

Clutch Friction and Torque Observer

6.1 Principle

During the design of the optimal engagement control the online least squares estimation of the characteristic curve $\Gamma_c(x_b)$ of the transmissible torque of the clutch as a function of the hydraulic actuator position x_b has been supposed perfect. This hypothesis allowed use of the clutch as a simple torque actuator thanks to the inversion of this characteristic curve done by the low levels of the AMT control strategy.

The $\Gamma_c(x_b)$ curve is the composition of two physically distinct components: the relation $F_n(x_b)$ between the normal force exerted on the friction pads and the hydraulic actuator position and $\Gamma_c(F_n)$ linking the normal force to the effective torque transmitted by the clutch. The former, mainly set by the washer and flat springs' stiffness, has quite a slow changing rate due to the friction pads wear and the aging of the springs. The second relation, on the other hand, is given by the friction-disk geometry and the friction coefficient that can have noticeable variations resulting from the heating of the friction surfaces. The least squares learning algorithm implemented in the low levels of the AMT strategy has as its main purpose the compensation of the slow changes in $F_n(x_b)$ due to the friction wear and squashing of the flat spring. The more rapid change of the friction coefficient due to the heating of the friction surfaces, weighted by the least squares algorithm, causes the $\Gamma_c(x_b)$ to be learned with a mean value of the friction coefficient γ.

The synchronization-assistance strategy, presented in Chapter 3, is particularly sensible to an error of the friction coefficient since the chosen optimal trajectory with its final conditions must be well adapted to the driveline real state and particularly the actual torque transmitted by the clutch.

In order to compensate this limitation of the AMT software lower levels two observers, one estimating the friction coefficient and one directly the clutch torque, have been developed. As will be presented in the next chapters these observers allow a better estimation of the actual transmitted clutch torque when activating the synchronization assistance.

6.2 Friction-coefficient Observer

6.2.1 Motivation

The torque control of the clutch is made possible by the inversion of the relation between the maximal transmissible torque by the clutch for a given position of the hydraulic actuator. This relation can be written as $\Gamma_c(x_b) = \gamma F_n(x_b)$ where $\gamma = 2\mu_d r_c$ is the coefficient linking the transmissible torque Γ_c to the normal force $F_n(x_b)$ exerted by the washer spring.

Figure 6.1. Schema of the clutch torque control detailing the washer-spring's characteristic $F_n(x_b)$ and the friction coefficient

Given the least squares estimation of the transmissible torque

$$\Gamma_c^{LSQ}(x_b) = \gamma^{LSQ} F_n^{LSQ}(x_b)$$

the actual transmitted clutch torque Γ_c, when asking for a target value $\bar{\Gamma}_c$, is

$$\Gamma_c = \gamma F_n \left(\left(\Gamma_c^{LSQ} \right)^{-1} \left(\bar{\Gamma}_c \right) \right) = \gamma F_n \left(\left(F_n^{LSQ} \right)^{-1} \left(\frac{\bar{\Gamma}_c}{\gamma^{LSQ}} \right) \right). \qquad (6.1)$$

Assuming a perfect estimation of the washer-spring characteristic

$$F_n^{LSQ}(x_b) \equiv F_n(x_b),$$

Equation 6.1 simplifies in

$$\Gamma_c = \frac{\gamma}{\gamma^{LSQ}} \bar{\Gamma}_c.$$

A correct estimation of the friction coefficient $\hat{\gamma}$ would allow, thus, a better control of the actually transmitted torque.

6.2.2 Driveline Models

The driveline simplified model in Equations 2.2a, 2.2b, 2.2c, and 2.2d together with the simplified friction model, valid for positive clutch sliding speeds $\omega'_e - \omega'_g$, gives the bilinear model

$$J_e \dot{\omega}'_e = \Gamma_e - \gamma F_n \tag{6.2}$$

$$J_g \dot{\omega}'_g = \gamma F_n - k'_t \theta' - \beta'_t \left(\omega'_g - \omega'_v \right) \tag{6.3}$$

$$J_v \dot{\omega}'_v = k'_t \theta' + \beta'_t \left(\omega'_g - \omega'_v \right) \tag{6.4}$$

$$\dot{\theta}' = \omega'_g - \omega'_v \tag{6.5}$$

$$\dot{\gamma} = 0, \tag{6.6}$$

having Γ_e and F_n as inputs. The friction coefficient γ figures as an additional constant state.

Since the clutch torque Γ_c given by the simplified friction model is independent of the clutch sliding speed, the previous model is actually composed by two separate parts: Equation 6.2 on one side and Equations 6.3, 6.4, and 6.5 on the other, sharing the additional Equation 6.6. Due to this separation the friction coefficient can be observed from the dynamics of only the section of the driveline upstream of the clutch, *i.e.* Equations 6.2 and 6.6.

6.2.3 Multi-input Multi-output Linear Time-variable Observer

The two bilinear systems, defined by the set of Equations (6.2)–(6.5) and (6.6), considering the complete driveline, or Equations 6.2 and 6.6, considering only the engine inertia, can also be thought of as parameter-controlled linear systems, *i.e.* systems in the form

$$\dot{x} = A(\vartheta)x + B(\vartheta)u$$
$$y = C(\vartheta)x,$$

where ϑ is a, possibly variable, parameter vector.

Different approaches [22, 4, 24, 6], globally known as *adaptive observers*, are available in the literature for the joint estimation of an eventually non-linear SISO system state and the parameters controlling its dynamic. These results have been extended to the MIMO case by Zhang in [37] and [38] whose main results are briefly presented here.

The system of interest in Zhang's paper is

$$\dot{x} = \mathbf{A}(t)x(t) + \mathbf{B}(t)u(t) + \Psi(t)\vartheta \tag{6.7a}$$

$$y(t) = \mathbf{C}(t)x(t), \tag{6.7b}$$

where $x(t) \in \mathbb{R}^n$, $u(t) \in \mathbb{R}^l$, $y(t) \in \mathbb{R}^m$ are respectively the state vector, the input and the output, $\mathbf{A}(t)$, $\mathbf{B}(t)$ and $\mathbf{C}(t)$ are known time varying matrices of appropriate size, $\vartheta \in \mathbb{R}^p$ a constant vector of unknown parameters, $\Psi(t) \in \mathbb{R}^n \times \mathbb{R}^p$ a known signal matrix. All the matrices $\mathbf{A}(t)$, $\mathbf{B}(t)$, $\mathbf{C}(t)$ and Ψ are supposed piecewise-continuous and uniformly bounded in time.

Under the following assumptions:

- There exists a matrix $\mathbf{K}(t) \in \mathbb{R}^n \times \mathbb{R}^m$ bounded in time such that

$$\dot{\eta}(t) = [\mathbf{A}(t) - \mathbf{K}(t)\mathbf{C}(t)]\,\eta(t) \tag{6.8}$$

  is globally exponentially stable; and

- Ψ is persistently exciting, *i.e.* given the signal matrix $\Upsilon(t) \in \mathbb{R}^n \times \mathbb{R}^p$ defined by the ODE system

$$\dot{\Upsilon}(t) = [\mathbf{A}(t) - \mathbf{K}(t)\mathbf{C}(t)]\,\Upsilon(t) + \Psi(t) \tag{6.9}$$

  there exist δ, T positive constants and a symmetric bounded positive matrix $\Sigma(t) \in \mathbb{R}^m \times \mathbb{R}^m$ such that for every t the following inequality holds true

$$\int_t^{t+T} \Upsilon^T(\tau)\mathbf{C}^T(t)\Sigma(t)\mathbf{C}(t)\Upsilon(\tau)\mathrm{d}\tau \geq \delta I.$$

For every positive symmetric matrix $\Gamma \in \mathbb{R}^p \times \mathbb{R}^p$, the system

$$\dot{\hat{x}}(t) = \mathbf{A}(t)\hat{x}(t) + \mathbf{B}(t)u(t) + \Psi(t)\hat{\vartheta}(t)$$
$$+ \left[\mathbf{K}(t) + \Upsilon(t)\Gamma\Upsilon^T(t)\mathbf{C}^T(t)\Sigma(t)\right][y(t) - \mathbf{C}(t)\hat{x}(t)] \tag{6.10}$$

$$\dot{\hat{\vartheta}}(t) = \Gamma\Upsilon^T(t)\mathbf{C}^T(t)\Sigma(t)\,[y(t) - \mathbf{C}(t)\hat{x}(t)] \tag{6.11}$$

is a globally exponentially stable adaptive observer.

The two previous bilinear systems can easily be expressed using a parameter-controlled linear system formulation in order to match Zhang's formulation in Equations 6.7a and 6.7b with

$$x = \begin{bmatrix} \omega_e' & \omega_g' & \omega_v' & \theta' \end{bmatrix}^T \quad u(t) = \Gamma_e(t) \tag{6.12a}$$

$$\Psi(t) = \Psi_M F_n(t) \quad \vartheta = \gamma \tag{6.12b}$$

$$\mathbf{A} = \begin{bmatrix} 0 & 0 & 0 & 0 \\ 0 & -\frac{\beta_t'}{J_g'} & \frac{\beta_t'}{J_g'} & -\frac{k_t'}{J_g'} \\ 0 & \frac{\beta_t'}{J_v'} & -\frac{\beta_t'}{J_v'} & \frac{k_t'}{J_v'} \\ 0 & 1 & -1 & 0 \end{bmatrix} \quad \mathbf{B} = \begin{bmatrix} \frac{1}{J_e} \\ 0 \\ 0 \\ 0 \end{bmatrix} \quad \Psi_M = \begin{bmatrix} -\frac{1}{J_e} \\ \frac{1}{J_g'} \\ 0 \\ 0 \end{bmatrix} \tag{6.12c}$$

$$\mathbf{C} = \begin{bmatrix} 1 & 0 & 0 & 0 \\ 0 & 1 & 0 & 0 \end{bmatrix} \tag{6.12d}$$

for the first system and

$$x = \omega'_e \quad u(t) = \Gamma_e(t) \quad \Psi(t) = \Psi_M F_n(t) \quad \vartheta = \gamma \tag{6.13a}$$

$$\mathbf{A} = 0 \quad \mathbf{B} = \frac{1}{J'_e} \quad \mathbf{C} = 1 \quad \Psi_M = -\frac{1}{J'_e} \tag{6.13b}$$

for the second.

In order to use the MIMO-LTV observer to jointly estimate the driveline state x and the friction coefficient γ the two assumptions have to be verified.

The first condition is easily verified since the couple $(\mathbf{A}, \mathbf{C})$ is observable for both systems, therefore a constant matrix $\mathbf{K}$ satisfying Equation 6.8 can be obtained by standard pole placement.

The following lemma is introduced to verify the second assumption.

Lemma 6.1. *Given the linear system*

$$\dot{x} = \mathbf{A}x + \mathbf{B}u \tag{6.14}$$

$$y = \mathbf{C}x \tag{6.15}$$

with $(\mathbf{A}, \mathbf{B})$ controllable and $(\mathbf{A}, \mathbf{C})$ observable, $y(t) = 0$ $\forall t \in T$ implies that $u(t) = 0$ over the same interval T.

Proof. $y(t) = 0$ over the interval T implies

$$\frac{d^k}{d^k t} y(t) = y^{(k)} = 0 \quad \forall k \in \mathbb{N}.$$

By definition of the observability matrix $\mathcal{O}$

$$\mathcal{O}x(t) = \left[y(t) \ y'(t) \cdots y^{(n-p)}(t) \right],$$

where $n = \text{rank}(A)$ and $p = \text{rank}(C)$. Since by hypothesis we have $\text{rank}(\mathcal{O}) = n$ then $x(t) = 0$ over the interval T.

By applying a similar line of reasoning on the controllability matrix $\mathcal{C}$ we have that $x(t) = 0$ over the interval T implies that $u(t) = 0$ for the same time interval.

$\square$

Defining $\upsilon = \mathbf{C}\Upsilon$ the persistently excited condition can be written as

$$\exists \, T, \delta \geq 0, \text{ such that } \int_t^{t+T} \upsilon^T \Sigma \upsilon \geq \delta I.$$

Since T and δ are arbitrary and the Σ matrix is symmetric and positive the previous condition is only satisfied if $\upsilon = 0$ over the interval T. Since the linear system is defined by the triplet $(\mathbf{A} - \mathbf{KC}), \Phi_M, \mathbf{C}$ is both controllable and observable, by virtue of the previously introduced lemma, the second assumption is verified for every F_n not identically equal to zero over the interval T.

The choice of the matrices K, Σ and Γ allows the convergence speed of the observer to be set.

6.2.4 Sampled-data Observer

The observer obtained from Equations 6.10, 6.11, and 6.9 by the substitution of Equations 6.12a, 6.12b, 6.12c, and 6.12d for the complete driveline model or Equations 6.13a and 6.13b for the driveline reduced to the engine mass is

$$\dot{\hat{x}} = \mathbf{A}\hat{x} + \mathbf{B}u + \hat{\gamma}\Psi_M F_n + \left[\mathbf{K} + \Upsilon\Gamma\Upsilon^T\mathbf{C}^T\Sigma\right][y - \mathbf{C}\hat{x}] \tag{6.16a}$$

$$\dot{\hat{\gamma}} = \Gamma\Upsilon^T\mathbf{C}^T\Sigma[y - \mathbf{C}\hat{x}] \tag{6.16b}$$

$$\dot{\Upsilon} = [\mathbf{A} - \mathbf{KC}]\Upsilon + \Psi_M F_n, \tag{6.16c}$$

where the Υ vector can be seen as a variable gain.

Since the AMT clutch and gearbox control unit is implemented through a digital computer the observer equations (6.16) must be sampled and transformed into a finite-difference system. Due to the variable gain and the bilinear nature of the system this step is not straightforward and a simple forward Euler approximation makes the system unstable.

The approximation method assuring the best level of stability is the bilinear or Tustin approximation

$$\frac{\mathrm{d}}{\mathrm{d}t}x \approx \frac{x(t_2) - x(t_1)}{2(t_2 - t_1)},$$

which for a dynamic system $\dot{x} = f(x, u)$ gives

$$x_2 - x_1 = \frac{\Delta t}{2}\left(f(x_2, u_2) + f(x_1, u_1)\right), \tag{6.17}$$

where Δt is the sampling interval.

The previous equation defines implicitly the new state vector sample x_2 as a function of the previous state vector sample x_1 and the input values u_1 and u_2 for the two sampling instants.

Compared to the Euler approximation defined by the relation

$$x_2 - x_1 = \Delta t f(x_1, u_1)$$

the Tustin approximation presents two difficulties: the solution of an implicit equation and the use of the value of the input u_2.

The analytic solution of the implicit equation (6.17) is theoretically possible with both the complete and the reduced driveline model, but in the former case the solution, obtained through the use of a computer algebra system (CAS) like Maple^{TM}, is too long to be practically useful. For the *simple* case the solution is over 200 symbols long and will not be reproduced here for practical reasons.

6.3 Clutch Torque Observer for Automated Manual Transmission

6.3.1 Principle

The friction-coefficient observer is based on the hypothesis of a good reconstruction of the washer-spring's characteristic $F_n(x_b)$ independently of the friction coefficient γ variation whose changes are mainly due to the heating of the friction surfaces. The heating of the friction disk also changes the flat spring stiffness and size, modifying the $F_n(x_b)$ curve especially around the contact point.

The friction-coefficient observer is highly sensitive to a wrong estimation of the contact point since a positive transmitted torque with an estimated zero normal force or the opposite case induce very strong variations in the estimated friction coefficient.

Since the F_n value is not always trustworthy the possibility of estimating the transmitted clutch torque without resorting to this signal, thanks to a class of observers called *unknown-input observers* [8], has been studied.

6.3.2 Unknown-input Observer

The basic idea of an unknown-input observer is the coupling of the dynamic model of the observed system with an autonomous model of the unknown-input. The prediction error feedback is used to guarantee the convergence of both the estimated state vector and the estimated system input.

The observed system, if we consider the reduced driveline model, is

$$J_e \dot{\omega}_e = \Gamma_e - \Gamma_c. \tag{6.18}$$

The engine control unit (ECU) gives an estimation of the mean output torque $\hat{\Gamma}_e$ based on several physical control parameters like, for example, the intake pressure, injected fuel quantity and the ignition point. The engine speed is measured by the ECU twice per revolution through an inductive sensor aimed at a toothed crown mounted on the flywheel. This measure, independently of the twice per revolution update, is broadcasted over the controller area network (CAN) to the gearbox control unit every hundredth of a second. This asynchronous setup causes about one repeated sample out of three. For simplicity this error is considered as a measurement noise ε_2; the system output is thus

$$\bar{\omega}_e = \omega_e + \varepsilon_2. \tag{6.19}$$

The chosen input model is a simple constant value, *i.e.*

$$\dot{\Gamma}_c = 0,$$

which gives, including the estimated unknown-input $\hat{\Gamma}_c$ as an additional state, the simple linear observer

$$J_e \dot{\hat{\omega}}_e = \hat{\Gamma}_e - \hat{\Gamma}_c + k_1(\bar{\omega}_e - \hat{\omega}_e) \tag{6.20a}$$

$$\dot{\hat{\Gamma}}_c = k_2(\bar{\omega}_e - \hat{\omega}_e). \tag{6.20b}$$

Since the observer is a simple linear system the sampling is straightforward.

6.3.3 Estimation Error Analysis

The main perturbations affecting the observer are a variation in the clutch transmitted torque, either due to a change of the normal force exerted on the friction surfaces or a change in the friction coefficient due to the heating of the friction surfaces, the previously described measurement noise on the engine speed due to asynchronicity between the engine control unit and the CAN bus updates and, finally, a possible estimation error on the engine output torque.

Considering the perturbed system

$$\dot{\omega}_e = \frac{1}{J_e'}(\Gamma_e - \Gamma_c) \tag{6.21a}$$

$$\dot{\Gamma}_c = \varepsilon_1 \tag{6.21b}$$

having as outputs

$$\bar{\omega}_e = \omega_e + \varepsilon_2$$

$$\hat{\Gamma}_e = \Gamma_e + \varepsilon_3,$$

where ε_1 represents a variation on the torque transmitted by the clutch, ε_2 the engine speed measurement noise and ε_3 the engine torque estimation error, and the corresponding unknown-input observer

$$\dot{\hat{\omega}}_e = \frac{1}{J_e'}\left(\hat{\Gamma}_e - \hat{\Gamma}_c\right) + k_1(\bar{\omega}_e - \hat{\omega}_e)$$

$$\dot{\hat{\Gamma}}_c = k_2(\bar{\omega}_e - \hat{\omega}_e),$$

the following theorem, whose demonstration is given in the Appendix B, applies. This theorem, based on the $\mathcal{L}_2$ gain of a linear system, allows to give an upper bound on either the joint estimation error, *i.e.* the quadratic norm of the vector formed by the engine speed and transmitted torque estimation error, or on the quadratic norm of the transmitted torque estimation error alone to be given.

Theorem 6.1 *Given the linear perturbed system*

$$\dot{z} = \boldsymbol{A}x + \boldsymbol{B}u + \boldsymbol{W}_1\epsilon_1$$
$$y = \boldsymbol{C}x + \boldsymbol{W}_2\epsilon_2,$$

with $(\boldsymbol{A}, \boldsymbol{C})$ *observable and a matrix* $\boldsymbol{K}$ *such that* $\boldsymbol{A} - \boldsymbol{K}\boldsymbol{C}$ *has real negative eigenvalues with linearly independent associated eigenvectors, the Luenberger observer*

$$\dot{\hat{x}} = \boldsymbol{A}\hat{x} + \boldsymbol{B}u + \boldsymbol{K}(x - \hat{x})$$
$$\hat{y} = \boldsymbol{C}\hat{x}$$

has an estimation error $\tilde{x} = x - \hat{x}$ *bounded by*

$$\|\tilde{x}\|_{\mathcal{L}_p} \leq \gamma_1\|\epsilon_1\|_{\mathcal{L}_p} + \gamma_2\|\epsilon_2\|_{\mathcal{L}_p} + \beta,$$

where

$$\gamma_1 = -\frac{\lambda_{max}}{\lambda_{min}^2}\|\mathbf{W}_1\|_2 \quad \gamma_2 = -\frac{\lambda_{max}}{\lambda_{min}^2}\|\mathbf{K}\mathbf{W}_2\|_2$$

$$\beta = \rho\|\tilde{x}(0)\|_2\sqrt{\frac{\lambda_{max}}{\lambda_{min}}}$$

$$\lambda_{max} = \max\{\lambda(\mathbf{A} - \mathbf{K}\mathbf{C})\}$$
$$\lambda_{min} = \min\{\lambda(\mathbf{A} - \mathbf{K}\mathbf{C})\}$$

$$\rho = \begin{cases} 1, & \text{if } p = \infty \\ \left(\frac{1}{p\lambda_{min}}\right)^{1/p}, & \text{if } p \in [1, \infty). \end{cases}$$

The joint estimation error

$$\tilde{x} = \begin{bmatrix} \omega_e - \hat{\omega}_e & \Gamma_c - \hat{\Gamma}_c \end{bmatrix}$$

is, thus, bounded by

$$\|\tilde{x}\|_{\mathcal{L}_2} \leq \gamma_1\left\|\begin{bmatrix} \varepsilon_1 \\ \varepsilon_3 \end{bmatrix}\right\|_{\mathcal{L}_2} + \gamma_2\|\begin{bmatrix} \varepsilon_2 \end{bmatrix}\|_{\mathcal{L}_2} + \beta,$$

where

$$\gamma_1 = -\frac{\lambda_{max}}{\lambda_{min}^2}\|\mathbf{B}_{o1}\|_2 = -\frac{\lambda_{max}}{J_e'\lambda_{min}^2}$$

$$\gamma_2 = -\frac{\lambda_{max}}{\lambda_{min}^2}\|\mathbf{B}_{o2}\|_2 = -\frac{\lambda_{max}}{\lambda_{min}^2}\sqrt{k_1^2 + k_2^2}$$

$$\beta = \rho\|\tilde{x}(0)\|_2\sqrt{\frac{\lambda_{max}}{\lambda_{min}}} = \|\tilde{x}(0)\|_2\sqrt{\frac{\lambda_{max}}{2\lambda_{min}^2}}$$

$$\lambda_{max} = \max\{\lambda(\mathbf{A}_o)\}$$
$$\lambda_{min} = \min\{\lambda(\mathbf{A}_o)\}$$

$$\mathbf{A}_o = \begin{bmatrix} -k_1 & -\frac{1}{J_e'} \\ -k_2 & 0 \end{bmatrix} \quad \mathbf{B}_{o1} = \begin{bmatrix} 0 & -\frac{1}{J_e'} \\ 1 & 0 \end{bmatrix} \quad \mathbf{B}_{o2} = \begin{bmatrix} -k_1 \\ -k_2 \end{bmatrix}.$$

In particular, considering just the transmitted torque estimation error $\tilde{T}_c$, we have

$$\|\tilde{T}_c\|_{\mathcal{L}_2} = \|C_o\tilde{x}\|_{\mathcal{L}_2} <= \|C_o\|_2\|\tilde{x}\|_{\mathcal{L}_2},$$

where

$$C_o = \begin{bmatrix} 0 & 1 \end{bmatrix}.$$

Since $\|C_o\|_2 = 1$ we have

$$\|\tilde{T}_c\|_2 \le \gamma_1\|\begin{bmatrix} \varepsilon_1 \\ \varepsilon_3 \end{bmatrix}\|_{\mathcal{L}_2} + \gamma_2\|\begin{bmatrix} \varepsilon_2 \end{bmatrix}\|_{\mathcal{L}_2} + \beta. \tag{6.24}$$

Since λ_{min} appears always in the denominator the estimation error norm is minimized for $\lambda_{max} = \lambda_{min} = \lambda$. In this case, γ_1, γ_2 and β can be further simplified to

$$\gamma_1 = -\frac{1}{J_e'\lambda} \quad \gamma_2 = -\frac{1}{\lambda}\sqrt{k_1^2 + k_2^2} \quad \beta = \|\tilde{x}(0)\|_2\sqrt{\frac{1}{2\lambda}}. \tag{6.25}$$

Since, for $\lambda_{max} = \lambda_{min} = \lambda$

$$\lambda = -\frac{k_1}{2}$$

and

$$k_2 = \frac{J_e'k_1^2}{4},$$

Equation 6.24 together with Equation 6.25, unsurprisingly, show that an extremely fast observer, i.e. when $\lambda \to \infty$, is very reactive and robust to input noise and initial error, i.e. $\gamma_1 \to 0$ and $\beta \to 0$, while being extremely subject to input noise, i.e. $\gamma_2 \to \infty$. This result could be used for optimal dimensioning of the observer given the error norms even if in practice the observer has been simply manually tuned for the best noise/fastness compromise.

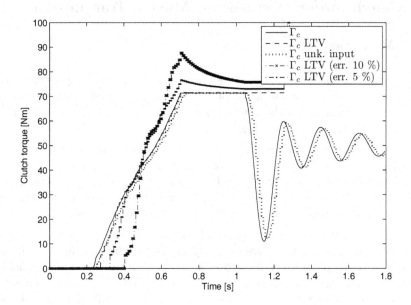

Figure 6.2. Results of the clutch torque estimation Γ_c for a simulated standing-start on flat ground for a Clio II 1.5dCi (K9K) using the unknown-input observer and the MIMO-LTV with $\Gamma_c = \hat{\gamma} F_n$ for the nominal case and for a 5% and 10% error on the contact point

6.3.4 Performance Comparison

Figure 6.2 allows a comparison of the performances of the MIMO-LTV observer and the unknown-input observer on a simulated standing-start on flat ground for a Clio II 1.5dCi (K9K). Due to the validity domain of the $\Gamma_c = \gamma F_n(x_b)$ relation and the persistent excitation assumption the MIMO-LTV observer is active only during the sliding phase. The unknown-input observer, on the other hand, is active during the whole simulation and after the engagement correctly follows the torque oscillation induced by the driveline.

In the nominal case, *i.e.* when the hypothesis $F_n(x_b) \equiv F_n^{LSQ}(x_b)$ of a good estimation of the normal force, the MIMO-LTV observer is well performing since it has no delay due to the input estimation. In the presence of a small error on the estimation of the washer-spring characteristic, on the other hand, the unknown-input observer guarantees a better estimation of the transmitted torque.

6.4 Clutch-torque Observer for Manual Transmission

6.4.1 Motivation

The changes of the F_n characteristic curve due to the heating of the friction disk have motivated the study of the unknown-input observer to obtain an accurate estimation of the transmitted clutch torque when activating the synchronization-assistance strategy.

The very good performances of this observer and its independence from the F_n estimation signal, usually only available on AMT vehicles, make it a possible candidate for advanced engine-control strategies during gearshifts. The main challenge in this kind of application is the convergence speed since the sliding phase of an upward gearshift is roughly of one tenth of a second.

This section is devoted to a refinement of the previous observer introducing a more detailed model of the driveline upstream of the clutch, a more realistic time-linear variation of the unknown-input and explicitly taking into consideration the non-uniform sampling of the engine speed. These improvements are needed to improve the convergence speed and to take full advantage of the implementation of the observer directly in the engine control unit instead of the gearbox control unit as was the case for the two previous observers. These refinements create a family of four observers whose performances will be tested with simulation and actual measures in order to find the best observer with respect to the convergence speed.

6.4.2 Observer Structure

The basic principle of the observer is the same, namely to couple a model of the part of the driveline upstream of the clutch with a model of the input evolution.

Two models of this part of the driveline have been considered: one simply composed by the engine mass J_e' accelerated by the engine torque Γ_e and braked by the clutch torque Γ_c, as in the previous case, and the other including a liberalized DMFW model as shown in Figure 6.3.

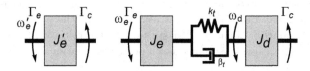

Figure 6.3. Model of the driveline part upstream of the clutch

The corresponding dynamic equations are

$$J'_e \dot{\omega}'_e = \Gamma_e - \Gamma_c \tag{6.26}$$

for the first one and

$$J_e \dot{\omega}_e = \Gamma_e - k_d \theta_d - \beta_d(\omega_e - \omega_g) \tag{6.27a}$$

$$J_d \dot{\omega}_g = k_d \theta_d + \beta_d(\omega_e - \omega_g) - \Gamma_c \tag{6.27b}$$

$$\dot{\theta}_d = \omega_e - \omega_g \tag{6.27c}$$

for the second. Equations (6.27) and (6.26) can be written in a standard state space representation

$$\dot{x} = \mathbf{A}x + \mathbf{B}u = \mathbf{A}x + \mathbf{B}_e \Gamma_e + \mathbf{B}_c \Gamma_c \tag{6.28a}$$

$$y = \omega_e = \mathbf{C}x, \tag{6.28b}$$

with

$$x = \begin{bmatrix} \omega'_e \end{bmatrix} \text{ or } x = \begin{bmatrix} \omega_e & \omega_d & \theta_d \end{bmatrix}^T \text{ and } u = \begin{bmatrix} \Gamma_e \\ \Gamma_c \end{bmatrix}.$$

Also, two models of the clutch torque evolution have been considered: either a constant value

$$\dot{\Gamma}_c = 0, \tag{6.29}$$

or a linear time variation

$$\dot{\Gamma}_c = \delta \tag{6.30a}$$

$$\dot{\delta} = 0. \tag{6.30b}$$

As before Equations 6.29 and 6.30 can be written in a standard state space representation

$$\dot{x}_c = \mathbf{A}_c x_c \tag{6.31a}$$

$$y_c = \Gamma_c = \mathbf{C}_c x_c, \tag{6.31b}$$

where

$$x_c = \begin{bmatrix} \Gamma_c \end{bmatrix} \text{ or } x = \begin{bmatrix} \Gamma_c & \delta \end{bmatrix}^T.$$

6.4.3 Continuous Unknown-input Observer

Coupling one of the two driveline models (Equations 6.27 or 6.26) with one of the two input evolution models (Equations 6.29 or 6.30) we obtain four models sharing the following common representation

$$\begin{cases} \dot{x}_{ob} = \begin{bmatrix} \mathbf{A} & \mathbf{B}_c \mathbf{C}_c \\ 0 & \mathbf{A}_c \end{bmatrix} x_{ob} + \mathbf{B}_e \Gamma_e = \mathbf{A}_{ob} x_{ob} + \mathbf{B}_e \Gamma_e \\ y_{ob} = \omega_e = \begin{bmatrix} \mathbf{C} & 0 \end{bmatrix} x_{ob} = \mathbf{C}_{ob} x_{ob} \end{cases} \tag{6.32}$$

where
$$x_{ob} = \begin{bmatrix} x & x_c \end{bmatrix}^T.$$

The Luenberger observer for the system (6.32), that is the unknown-input observer with a constant or linearly time variable unknown-input hypothesis, is

$$\dot{\hat{x}}_{ob} = \mathbf{A}_{ob}\hat{x}_{ob} + \mathbf{B}_e\Gamma_e + \mathbf{K}(y_{ob} - \hat{y}_{ob}) \qquad (6.33)$$

$$= (\mathbf{A}_{ob} - \mathbf{C}_{ob}\mathbf{K})\hat{x}_{ob} + \begin{bmatrix} \mathbf{B}_e \\ \mathbf{K} \end{bmatrix} \begin{bmatrix} \Gamma_e & y_{ob} \end{bmatrix}$$

$$= \mathbf{A}_{bf}\hat{x}_{ob} + \mathbf{B}_{bf} \begin{bmatrix} \Gamma_e & y_{ob} \end{bmatrix}, \qquad (6.34)$$

where $\mathbf{K}$ is a gain matrix such that the estimation error dynamic $x_{ob} - \hat{x}_{ob}$ is stable.

Since the structure of the observer is not changed the estimation error analysis and bounding given in the previous section applies to the family of observers expressed by the previous equations.

6.4.4 Non-uniform Sampling

As previously highlighted, the main challenge to face for the utilization of an unknown-input observer to estimate the transmitted clutch torque during a gearshift is the convergence speed since the sliding phase of a gearshift is very short.

The standard engine speed measure, refreshed twice per crankshaft revolution, is not enough for our purposes: for example at 1800 rpm the sampling frequency is 60 Hz meaning that just 6 samples will be available during the sliding phase. A second measure of the engine speed updated twelve times per revolution is available as an ECU internal signal generated by the passage of a group of five teeth out in front of the inductive sensor. Figure 6.4 shows a comparison between the standard engine measure, the engine speed measure given by packets of five teeth and a tooth by tooth engine speed measure not available on a standard car. Oscillations in the engine speed are due to the engine acyclicity, *i.e.* the oscillation in the instantaneous engine output torque induced by the impulsive nature of the explosions in an internal combustion engine.

The observer code has to be executed at each update of the engine speed signal, *i.e.* every passage of a packet of five teeth in front of the inductive sensor. Between one TDC and the other the engine torque is supposed constant. For an engine speed of ω_{e0}, under the hypothesis of a constant engine speed, the next update will happen in $\Delta t = \frac{\pi}{6\omega_{e0}}$ s. Since the sampling time is a function of the system state the equivalent sampled system cannot be obtained in the usual way.

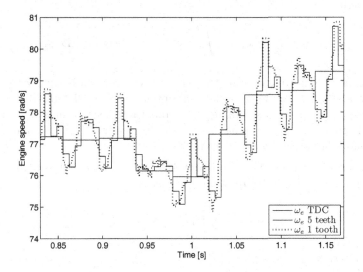

Figure 6.4. Comparison between different engine-speed measurements: the solid line is a standard twice-per-revolution measure over 30 teeth, the dashed line is a twelve-times-per-revolution measure over 5 teeth and the dotted line is a direct tooth-by-tooth measure obtained through an external 20 kHz sampling

Exact Sampling

Given the linear system

$$\dot{x} = \mathbf{A}x + \mathbf{B}u \tag{6.35}$$

$$y = \mathbf{C}x \tag{6.36}$$

the state x evolution over the time interval Δt is

$$x(t + \Delta t) = \mathrm{e}^{A\Delta t}x(t) + \int_{t}^{t+\Delta t} \mathrm{e}^{A(\Delta t - \tau - t)} Bu(\tau)\mathrm{d}\tau.$$

If the signals $x(t)$ and $u(t)$ are sampled with a non-uniform sampling time Δt_k using a reconstructive signal $\psi_k(t)$, *i.e.* for each sampling interval the sample x_k gives the amplitude of the $\psi_k(t)$-shaped function:

$$x(t) = \sum_{k=-\infty}^{\infty} \psi_k \left(t - \Delta T_k\right) x_k$$

$$u(t) = \sum_{k=-\infty}^{\infty} \psi_k \left(t - \Delta T_k\right) u_k, \quad \Delta T_k = \sum_{j=-\infty}^{k} \Delta t_j,$$

we can write

$$x_{k+1} = e^{\mathbf{A}\Delta t_k} x_k + \left[\int_0^{\Delta t_k} e^{\mathbf{A}(\Delta t_k - \tau)} \psi_k(\tau) d\tau \right] \mathbf{B} u_k. \tag{6.37}$$

In the case of uniform sampling the reconstructive signal is simply the same function for every sample subject to an opportune time shifting. In order to cope with a non-uniform sampling the reconstructive signal must not only be shifted in time to the beginning of the sampling interval but also matched to the variable sampling interval length Δt_k.

Using a zero-order hold the ψ_k function is simply a rectangular function over the interval $[t_k, t_k + \Delta t_k]$; Equation 6.37 is then

$$x_{k+1} = e^{\mathbf{A}\Delta t_k} x_k + \left[\int_0^{\Delta t_k} e^{\mathbf{A}(\Delta t_k - \tau)} d\tau \right] \mathbf{B} u_k. \tag{6.38}$$

Defining

$$\mathbf{A}_d(\Delta t) = e^{\mathbf{A}\Delta t} \tag{6.39a}$$

$$\mathbf{B}_d(\Delta t) = \left[\int_0^{\Delta t} e^{\mathbf{A}(\Delta t - \tau)} d\tau \right] \mathbf{B} \tag{6.39b}$$

$$\mathbf{C}_d(\Delta t) = \mathbf{C}, \tag{6.39c}$$

we have, finally, the sought for finite-difference equation

$$x_{k+1} = \mathbf{A}_d(\Delta t_k) x_k + \mathbf{B}(\Delta t_k) u_k$$
$$y_k = \mathbf{C}_d(\Delta t_k) x_k.$$

Furthermore, if the $\mathbf{A}$ matrix is invertible, Equation 6.39b can be written as

$$\mathbf{B}_d(\Delta t) = \mathbf{A}^{-1} \left(e^{\mathbf{A}\Delta t} - \mathbf{I} \right) \mathbf{B}. \tag{6.40}$$

Approximated Sampling

It would be practical to avoid having to calculate Equation 6.39a, 6.40, and 6.39c or having to store their values for every possible value of Δt. To solve this difficulty the matrix exponential is approximated with its Taylor series truncated at the second order. This operation is exactly equivalent to a Euler approximation of its dynamic equation and has been preferred to a bilinear or Tustin approximation since it avoids a one-sample delay needed to solve the implicit equation. The comparison with the exact solution shows that the approximation error is negligible for a non-uniform sampling rate of twelve samples per revolution.

$$e^{\mathbf{A}\Delta t} = \mathbf{I} + \mathbf{A}\Delta t + o(\Delta t^2). \tag{6.41}$$

Substituting Equation 6.41 in Equations 6.39a, 6.40 and, (6.39c) we obtain

$$\mathbf{A}_d(\Delta t) = \mathbf{I} + \mathbf{A}\Delta t + o(\Delta t^2) \tag{6.42a}$$

$$\mathbf{B}_d(\Delta t) = \Delta t \mathbf{B} + o(\Delta t^2) \tag{6.42b}$$

$$\mathbf{C}_d(\Delta t) = \mathbf{C}. \tag{6.42c}$$

Exact Sampled Observer

Applying the exact sampling to the continuous observer (6.34) we have

$$\hat{x}_{o_{k+1}} = \mathbf{A}^d_{bf_k}\hat{x}_{o_k} + \mathbf{B}^d_{bf_k}\left[\Gamma_e \; y_o\right]^T \tag{6.43}$$

$$\mathbf{A}^d_{bf_k} = e^{\mathbf{A}_{bf}\Delta t_k}$$

$$\mathbf{B}^d_{bf_k} = \mathbf{A}^{-1}_{bf}\left(e^{\mathbf{A}_{bf}\Delta t_k} - \mathbf{I}\right)\mathbf{B}_{bf}$$

$$\Delta t_k = \frac{\pi}{6\omega_{e_k}},$$

where $\mathbf{A}_{bf}$ and $\mathbf{B}_{bf}$ depend on the chosen driveline and input model. At each step the matrices defining the finite difference system are recalculated on the base of the current engine speed.

This solution, clearly inpractical, has as its only purpose to be a reference to measure the approximation error.

Approximated Sampled Observer

Applying the approximated sampling to (6.34) we have

$$\hat{x}_{o_{k+1}} = (\mathbf{I} + \mathbf{A}_{bf}\Delta t_k)\,\hat{x}_k + \Delta t_k \mathbf{B}_{bf}\left[\Gamma_e \; y_o\right]^T + o\left(\Delta t_k^2\right) \tag{6.44}$$

$$\approx (\mathbf{I} + \mathbf{A}_{bf}\Delta t_k)\,\hat{x}_k + \Delta t_k \mathbf{B}_{bf}\left[\Gamma_e \; y_o\right]^T \tag{6.45}$$

$$\Delta t_k = \frac{\pi}{6\omega_{e_k}}.$$

In the case of the approximated sampling observer no complex matrix calculation is required on-line and the observer is reduced to a non-linear finite-difference system with a number of equations between 2 and 5 depending on the chosen models.

6.4.5 Results

The four possible sampled observers have been tested on a 1-2 upshift both in simulation and with actual data captured on a Megane II 2.0 gasoline (F4R). Results show that:

- The error induced by the Euler approximation is negligible even for rapid gearshifts (Figure 6.5) since twelve samples per revolution induce a short enough sampling time.

- The transmitted-torque estimation signal is not noisy when using the captured data despite a good convergence speed.

- In simulation, the convergence speed is adapted even to very fast engagements. Figure 6.6 shows an estimated torque close to 70% of its actual value at synchronization. As the actual transmitted torque is unknown on a real vehicle a similar test cannot be performed with captured data.

- The observer providing the best performance is given by the combination of the driveline model without DMFW and a linear clutch torque evolution. Since the standard profile of the clutch pedal position trajectory for an upshift is quite close to a simple linear time variation the fact that a linear-variation hypothesis performs better is understandable. On the other hand the fact that a more refined model of the driveline upstream of the clutch induces worse performances is somewhat surprising but can probably be explained by the fact that two additional, non-measured, states have to be observed. It might be noted, though, that the observer using a more refined driveline model, while slower, does not overshoot the actual transmitted torque, as shown in Figure 6.6.

Since no measure of the actual transmitted torque was available for track-test measures[1] the results shown in Figure 6.5 testify mainly to the observer's robustness to noise measurements and the validity of Euler's approximation since no difference can be seen between the exact and approximated sampling. Results obtained on a simulation run of a very fast engagement instead, shown in Figure 6.6 allow comparison of the estimated toque against the actual transmitted torque, showing a sufficiently high convergence speed.

6.5 Conclusions

In this chapter two main observers have been presented: a friction-coefficient observer based on a multi-input multi-output linear time-variant (MIMO-LTV) extension of the class of observers known as *adaptive observers* for

---

[1] The actual value cannot be directly easily measured but can be estimated using torque sensors on the transmission shafts. Unfortunately, such a measure was not available on the test vehicle.

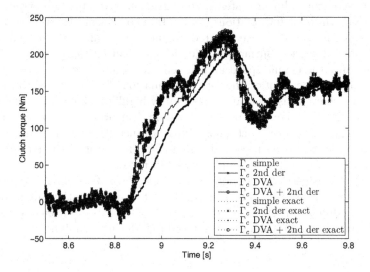

Figure 6.5. Results of the clutch torque observers for a 1-2 upshift using actual data measured on a Megane II 2.0 essence (F4R)

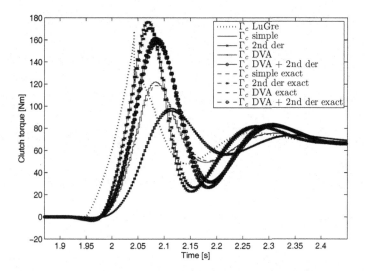

Figure 6.6. Results of the clutch torque observers for a simulated fast 1-2 upshift with a sliding time of just 0.05 s

parameter-controlled linear systems, and a transmitted torque observer based on an unknown-input formulation. The friction-coefficient observer has as inputs the engine speed, the estimated engine torque and the normal force exerted on the friction surfaces, while the transmitted-torque observer has as inputs only the engine speed and the engine torque. The first type of observer is highly dependent on the correct estimation of the washer spring's and flat spring's characteristics that, as previously highlighted, can change due to the friction heating. For this reason in the implementation, presented in the following chapter, of the synchronization-assistance strategy the transmitted torque observer has been preferred. Finally, a refinement of the transmitted torque observer, meant to be implemented in the engine control unit, taking explicitly in consideration the non-uniform sampling of the engine-speed has been analyzed and shown to be sufficiently fast to estimate the clutch torque during gearshifts and, thus, to be the base for an advanced engine torque control strategy for MT aiming to reduce lurch during gearshifts.

7

Experimental Results and Control Evaluation

7.1 Track Testing

This research work has initially been focused toward a comprehension and an improvement in MT clutch-related driving comfort. Since the simple mechanical solutions initially proposed have not been deemed sufficient, attention has been put to more complex active systems comprising an actuator acting directly on the clutch's fingers allowing at least a partial decoupling from the clutch-pedal position.

The proposed strategies should have been tested, normally, on a clutch-by-wire vehicle. Since a vehicle with this kind of transmission was not available for track testing, the strategies have been implemented on a programmable AMT prototype, lacking the clutch pedal. Experimental results have shown the pertinence of the ideal synchronization condition and the subjective comfort induced by the synchronization assistance strategies. The lack of a clutch pedal, on the other hand, made it impossible to verify the influence of the assistance and particularly the lack of any lurch at synchronization on the driver's behavior.

In the following sections of this chapter we will detail the implementation of the synchronization-assistance strategy and its experimental results obtained on a Clio II AMT prototype together with the current work for including an unknown-input observer in the new generation of engine-control software. All the track tests have been performed on the premises of the Centre Technique Renault of Lardy situated in the suburbs of Paris.

7.2 Synchronization-assistance Strategy

7.2.1 Clio II K9K Prototype

The vehicle used for testing the synchronization assistance strategy is a Clio II 1.5dCi prototype equipped with a five-speed Renault JH gearbox controlled by a hydraulic AMT module by Magneti Marelli. The AMT module, which also controls the clutch engagement and disengagement, is not driven by an industrial computer placed under the hood, as in the standard configuration, but by a dSpace® rapid prototyping card part of an IBM™compatible PC situated in the booth.

Figure 7.1. Clio II 1.5dCi prototype vehicle on the parking lot of the testing track of Centre Technique Renault in Lardy

The Figure dSpace® shows the connection diagram of the rapid prototyping systems installed on the vehicle. The whole gearbox control strategy is programmed in MATLAB® thanks to its various extensions like Simulink® for continuous block diagrams, Stateflow® for finite state machines or the C language for the external s-functions blocks. This program is first completely

---

[1] Minimal test weight including vehicle fluids, control and recoding equipment and the driver.

Figure 7.2. View of the cockpit, on the right the screen and the keyboard connected to the IBM^{TM}computer in the booth can be seen

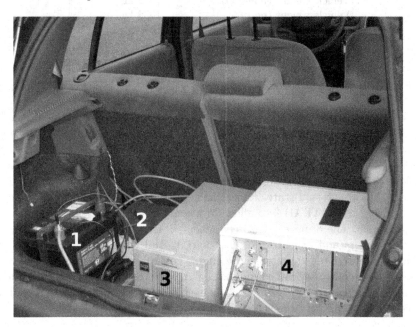

Figure 7.3. View of the rapid prototyping system placed in the booth. 1 Lead-acid battery powering the IBM^{TM}compatible PC and the interface rack 2 General switch 3 IBM^{TM}compatible PC 4 interface rack

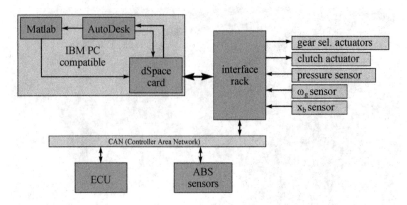

Figure 7.4. Simplified connection diagram of the dSpace® system

Table 7.1. Technical data of the Clio II AMT prototype

Body type	5 doors (B65)
Engine	1.5 dCi turbo diesel (K9K)
Engine displacement	1461 cm^3
Power	70 kW
Vehicle weight (empty)	990 kg
Vehicle weight (nominal)[1]	1212 kg
Vehicle weight (max)	1535 kg
J_e'	0.158 kgm^2
J_g'	$6.53 \cdot 10^{-3}$ kgm^2
Mean clutch radius r_c	93.325 mm
Wheel radius R_w	0.289 m

translated in C code, compiled and loaded on the DSP controlling the dSpace® card. Once the program has been loaded the card itself is completely independent, but for the energy supply, from the hosting computer. The AutoDesk® application is a graphical interface running on the IBM^{TM}PC allowing interaction with the loaded controller either changing its internal control variables or recoding its internal states. Once the capture is completed, these data can be exported to MATLAB® to allow for a more refined analysis of the system performance. The dSpace® card is connected, through the interface rack, to all the sensors and actuators of the gearbox and to the CAN bus through which it can exchange data with the engine control unit and the ABS sensors. The gearbox controller can, through the CAN interface, issue a torque request to the engine control unit but this request is not guaranteed to be honored since it has no priority over the ECU internal strategies.

7.2.2 Control Sequencing

The gearbox controller found on the prototype vehicle is a Renault proprietary software completely independent from the standard Magneti Marelli controller usually found on these cars.

Seen as a whole the AMT controller is a hybrid system, *i.e.* having both continuous and discrete states.[1] The various states of the AMT transmission (e.g. standing-start, gearshift, selected speed, neutral, *etc.*) are all part of a module assuring the nominal mode. Each state has a dedicated controlling module that is activated and deactivated when needed by the nominal mode supervisor. Figure 7.5 shows an idealized schema of this control structure, the actual interconnection is somewhat more complex due to the presence of a shared input-output layer and the close interconnection between the nominal mode and the diagnostic and security modules.

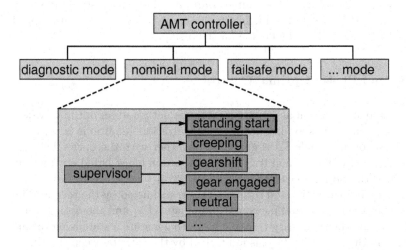

Figure 7.5. Idealized representation of the AMT gearbox controller internal structure. When in nominal mode the supervisor manages the activation of the various modules.

The actual implementation of the synchronization-assistance strategy, detailed in the following sections, translates to the introduction of a new standing-start module. The internal sequencing of the standing-start is controlled by the flow chart shown in Figure 7.6.

---

[1] Since the actual implementation is a program running on a finite state machine the controller itself cannot be but a finite state machine with discrete states; for clarity reasons we will introduce a distinction between the continuous sampled and quantized dynamics and the sequencing parts.

Figure 7.6. Sequencing flow chart for the synchronization assistance strategy

7.2.3 First Phase: Open-loop Control

When the standing-start module, including the synchronization-assistance strategy, is activated the strategy itself is kept in the idle state till the clutch sliding speed y_1 is lower than the activation threshold y_{act}. During this phase the clutch torque is controlled through simple open-loop tables using as only input the throttle pedal position x_p. The main focus of this phase is assuring a short engagement time. In the implementation used for track testing the clutch torque profiles were simple saturated ramps with the slope and the final value function of x_p. This simple solution, adapted for standing-starts on flat or slightly sloped ground, has to be upgraded with an engine-speed feedback to avoid stalling the engine on steep slopes.

7.2.4 Second Phase: Optimal Control

When the synchronization-assistance strategy is activated an optimal trajectory reaching the corresponding ideal synchronization conditions is selected. The optimal torque profile Γ_c^* is generated together with the optimal sliding speed profile y_1^* by the optimal trajectory-generation block based on the time passed since the activation of the optimal control $t - t_0$, the initial engine torque $\Gamma_e(t_0)$ and the estimated torque $\Gamma_c(t_0)$ transmitted by the torque. The difference between the actual measured sliding speed y_1 and the optimal profile y_1^* is used by the trajectory-tracking block to generate a stabilizing torque Γ_{stab}. Finally the sum $\Gamma_c^* + \Gamma_{stab}$ is multiplied by the ratio between the clutch torque target and the actual estimated transmitted torque at activation instant t_0 in order to compensate for a possible gain error on the washer-spring estimated characteristic $\Gamma_c^{LSQ}(x_b)$ used by the gearbox control lower levels. The transmitted clutch torque estimation is not used directly as a feedback since the convergence speed of the unknown-input observer implemented in the AMT controller is not sufficient, while, thanks to the saturated ramp profile used in the open-loop phase, the transmitted torque estimation at t_0 is quite precise.

In the following paragraphs we will analyze in some detail the actual implementation of the trajectory-generation and the trajectory-stabilization blocks.

Trajectory Generation

The two solutions of the finite-time optimal control problem detailed in Chapter 3 are not easily implementable online.

The dynamic Lagrangian multipliers method is limited by the bad conditioning of the φ_{12} matrix; the need to resort to a variable-precision arithmetic makes the computational cost associated with the solution of the linear problem (4.21) too high.

The solution using a QP formulation without any further simplification such as that presented in the Chapter 4 is also computationally too expensive to be used for an online solution.

In order to allow an offline computation of the optimal trajectories a simplifying assumption has been made. Under the hypothesis of lack of driveline oscillations at synchronization instant t_0, i.e. $y_2(t_0) = 0$, by using the dynamic equations of the simplified system and the definition of the threshold speed (4.23) the initial state vector results

$$x(t_0) = \begin{bmatrix} y_1(t_0) \\ y_2(t_0) \\ \theta(t_0) \\ \Gamma_c(t_0) \end{bmatrix} = \begin{bmatrix} \alpha(t_s - t_0)\left(\frac{\Gamma_e(t_0)}{J'_e} - \frac{J'_e+J'_g+J'_v}{J'_e(J'_g+J'_v)}\Gamma_c(t_0)\right) \\ 0 \\ \frac{1}{k'_t}\frac{J'_v}{J'_g+J'_v}\Gamma_c(t_0) \\ \Gamma_c(t_0) \end{bmatrix} \tag{7.1}$$

$$= \mathcal{F}_0(\Gamma_e(t_0), \Gamma_c(t_0)), \tag{7.2}$$

where $\alpha \in (0,1)$ is a coefficient shaping the optimal trajectory. The corresponding ideal synchronization conditions are

$$x(t_s) = \begin{bmatrix} y_1(t_s) \\ y_2(t_s) \\ \theta(t_s) \\ \Gamma_c(t_s) \end{bmatrix} = \begin{bmatrix} 0 \\ 0 \\ \frac{1}{k'_t}\frac{J'_v}{J'_e+J'_g+J'_v}\Gamma_e(t_0) \\ \frac{J'_g+J'_v}{J'_e+J'_g+J'_v}\Gamma_e(t_0) \end{bmatrix} = \mathcal{F}_s(\Gamma_e(t_0), \Gamma_c(t_0)). \tag{7.3}$$

Since the optimal trajectory is defined by the initial and final state vectors, which are functions only of $\Gamma_e(t_0)$ and $\Gamma_c(t_0)$, we can define

$$\Gamma_c^*(t) = \mathcal{F}_c(t, \Gamma_e(t_0), \Gamma_c(t_0)) \tag{7.4}$$

$$y_1^*(t) = \mathcal{F}_y(t, \Gamma_e(t_0), \Gamma_c(t_0)). \tag{7.5}$$

The offline calculation of a battery of optimal trajectories for several values of $\Gamma_e(t_0)$ and $\Gamma_c(t_0)$ allows sampling of the functions (7.4) and (7.5) and storing their values in a $3D$ look-up table; thus a series of optimal trajectories for different values of initial engine and clutch torque are available. The actual values of the optimal clutch torque or of the sliding speed are obtained online through a simple $2D+1$ interpolation between the closest optimal trajectories; a real $3D$ interpolation not being necessary since, due to the sampled nature of the controller, an exact value for the time dimension is always available.

As shown in Figures 4.11 and 4.12 the optimal trajectories for $\alpha = 0.5$ are quite close to a straight line between $\Gamma_c(t_0)$ and $\Gamma_c(t_f)$. The optimal trajectories can thus be also approximated by

$$
\Gamma_c^* = \Gamma_c(t_0) + \frac{\Gamma_c(t_s) - \Gamma_c(t_0)}{t_s - t_0}(t - t_0)
$$

$$
y_1^* = y_1(t_0) + \frac{\Gamma_e(t_0)}{J_e'}(t - t_0)
$$

$$
- \frac{J_e' + J_g' + J_v'}{J_e'(J_g' + J_v')}\left(\Gamma_c(t_0)(t - t_0) + \frac{\Gamma_c(t_s) - \Gamma_c(t_0)}{(t_s - t_0)}\frac{(t - t_0)^2}{2}\right),
$$

where y_{10} is defined by Equation 4.23 with $\alpha = 0.5$. This sub-optimal approximation does not satisfy the ideal synchronization conditions but when implemented and tested on the test vehicle delivers a subjective level of comfort comparable with the optimal solution while avoiding the complexity involved in an offline optimal trajectory computation and online interpolation.

Trajectory Tracking

The trajectory stabilization has two main difficulties: an actuator with a limited bandwidth and the presence of auto-induced thermo-elastic vibrations called TEI (thermo-elastic instability) in technical papers and hot judder in the automotive industry.

These oscillations are due to the interaction between the friction heating, the thermal deformation and the stiffness of the friction surfaces. This effect, first studied by Barber in 1969 [3], leads to the formation of hot spots on the friction surfaces and mechanical vibrations that can sensibly reduce the performances and endurance of the systems. The friction frequency depends on the friction surfaces characteristics and the connected masses; the clutch of the test vehicle presents a TEI oscillation, easily identifiable in the first part of Figure 7.7, having a frequency for the first gear of about 15 Hz. The higher-frequency oscillation visible after the engagement is due to the engine acyclicity that is visible on a plot of the sliding speed due to the higher sampling rate used for the gearbox speed. This research did not further analyze this effect, the interested reader can find more detailed analysis in the ample literature on the subject [2, 36, 35].

The hydraulic clutch actuator has been identified together with its positioning control loop and can be modeled by a second-order system with a cut-off frequency of 19 Hz and a damping coefficient of $\zeta = 0.7$. This bandwidth does not allow for an active damping of the TEI oscillations and any attempt in

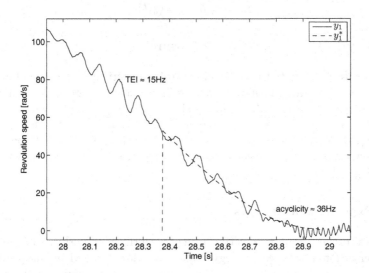

Figure 7.7. Example of trajectory tracking, the measured sliding speed is drawn as a solid line, while the optimal trajectory is dashed. A TEI oscillation is visible before the synchronization, while after it the higher-frequency engine acyclicity is present.

this direction, anyway, would be greatly challenged by the highly non-linear nature of the phenomenon.

The controller

$$\frac{\Gamma_{stab}(s)}{(y_1 - y_1^*)(s)} = \frac{0.1997s^2 - 0.1845s + 0.1327}{s^2 - 1.5024s + 0.5359}$$

has been obtained by direct synthesis imposing a closed-loop damped pole at 5 Hz, while keeping the natural driveline oscillation mode to reduce the control activity.

The clutch sliding speed is measured by the difference between the engine speed ω_e' and the gearbox primary shaft speed ω_g'. The latter is measured directly by the AMT controller through an inductive sensor aimed at the rear speed gear; the former, instead, is measured by the engine control unit and broadcasted on the CAN. This arrangement induces a delay of about 0.04 s and a possible lack of update due to the asynchronism between the engine control unit and the CAN. In order to limit the effect of these two perturbations on the trajectory tracking a simple state observer

$$\hat{\omega}_e(t) \approx \hat{\omega}_e(t - \Delta t) + \frac{1}{J_e'} \left[\bar{\Gamma}_e(t - \Delta t) - \hat{\Gamma}_c(t - \Delta t) \right] \Delta t$$

has been integrated to the clutch torque unknown-input observer. The measured clutch sliding speed is finally calculated as

$$y_1(t) = \hat{\omega}_e(t) - \omega'_g(t).$$

7.2.5 Third Phase: Final Clutch Closure

After an interval $t_s - t_0$ the optimal control is deactivated and the clutch is fully closed in a two-part movement: initially with a progressive closing to reduce to zero the eventual remaining sliding speed and finally with a rapid movement. This phase is never completely executed since half-way through the fast final closing the supervisor detects the end of the standing-start operation and deactivates the synchronization-assistance strategy and switches to the selected speed mode.

7.2.6 Experimental Results

The track testing has allowed verification of both the good performances of the synchronization-assistance strategy implementation detailed in the previous sections and the effective driving comfort improvement perceived by the driver thanks to the use of this strategy.

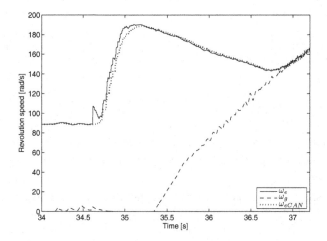

Figure 7.8. Capture of a standing-start on flat ground of the Clio II AMT prototype showcasing the synchronization-assistance strategy. The optimal control is activated at 36.4 s and leads to a synchronization at 36.9 s slightly in advance over the optimal 37 s value.

Figures 7.8–7.11 show the measures relative to a standing-start on flat ground with a nominal vehicle weight and a medium throttle position having an engagement time of about two seconds. Just after 36.4 s the synchronization-assistance strategy is activated leading to a partial reopening of the clutch

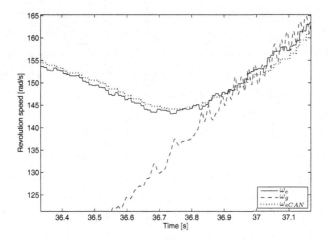

Figure 7.9. Detail of Figure 7.8 highlighting the effect of partial clutch reopening

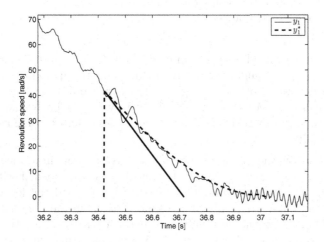

Figure 7.10. Clutch sliding speed relative to the standing-start operation shown in Figure 7.8. The thin solid line is the measured trajectory, the dashed line is the optimal reference trajectory and the black bold straight line shows the profile the sliding speed would have followed if no synchronization assistance was present.

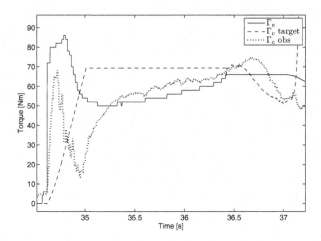

Figure 7.11. Engine torque, requested clutch torque and estimated clutch torque relative to the standing-start operation in Figure 7.8. The partial clutch reopening reduces the transmitted torque from its initial value of almost 70 Nm to about 50 Nm with an engine torque of 66 Nm.

reducing the transmitted torque from its initial value of almost 70 Nm to just about 50 Nm with an engine torque of 66 Nm. This torque reduction assures a zero time derivative of the sliding speed at synchronization together with the ideal synchronization conditions. Subjectively, this is felt by the driver as a smooth transition from the sliding to the engaged phases, while assuring a brilliant standing-start.

In order to verify the robustness of the control system implementing the synchronization-assistance strategy the vehicle has been loaded with a 400 kg ballast load reaching a total mass of 1612 kg, beyond the legal loading limit. The standing-start has been as comfortable as in the nominal case, as shown in Figure 7.12, the only difference being a longer sliding time during the open-loop phase visible in Figure 7.13 as a lower time derivative of the sliding speed prior to the assistance activation.

Finally, as previously briefly presented in the trajectory generation paragraph of Section 7.2.4, for a particular choice of the clutch sliding speed threshold triggering the activation of the synchronization-assistance strategy the clutch torque optimal trajectory can be approximated by a simple linear variation between the initial and the final condition. Results in Figures 7.15–7.17 show a level of perceived comfort equal to that obtained with a fully fledged optimal trajectory. In this example a change in the clutch characteristic induces quite a strong underestimation of the actual transmitted torque since the target value is lower than the engine torque and yet the engine speed is decreasing. The static gain correction obtained using the transmitted torque estimation

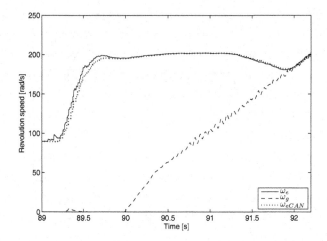

Figure 7.12. Capture of a standing-start on flat ground of the Clio II AMT prototype showcasing the synchronization-assistance strategy robustness under a 400 kg ballast load

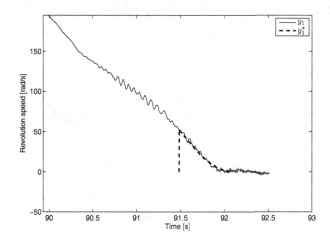

Figure 7.13. Clutch sliding speed relative to the standing-start operation shown in Figure 7.12. Due to the ballast load the derivative of the sliding speed is noticeably lower than the forecasted one at the beginning of the optimal trajectory.

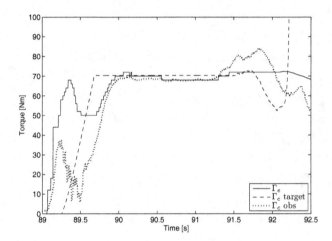

Figure 7.14. Engine torque, requested clutch torque and estimated clutch torque relative to the standing-start operation in Figure 7.12

given by the unknown-input observer corrects this error and allows a correct synchronization.

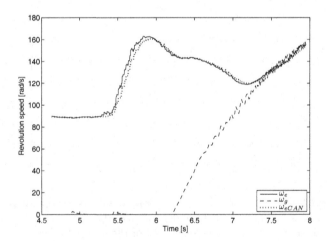

Figure 7.15. Capture of a standing-start on flat ground of the Clio II AMT prototype showcasing the sub-optimal synchronization-assistance strategy

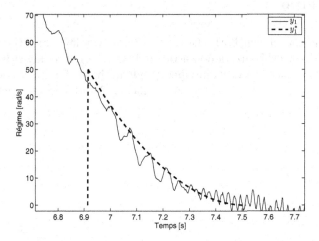

Figure 7.16. Clutch sliding speed relative to the standing-start operation shown in Figure 7.15. Due to the linear decrease of the clutch torque the sub-optimal reference trajectory is simply an arc of a parabola.

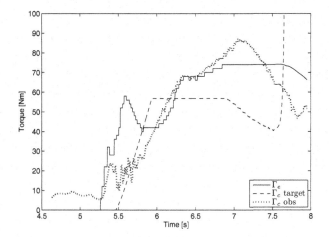

Figure 7.17. Engine torque, requested clutch torque and estimated clutch torque relative to the standing-start operation in Figure 7.15. The linear decrease of torque is clearly visible at the end of the engagement. It can also be noted that, due to a change of the clutch characteristic, while the target value of Γ_c is lower than Γ_e, the value estimated by the unknown-input observer is higher than the engine torque, which is consistent with the decrease of engine speed shown in Figure 7.15.

7.3 Conclusions

Track testing has shown both the excellent comfort level and the good robustness of the synchronization-assistance strategy for standing-starts on flat
ground. The absence of any engine speed feedback in the first open-loop phase,
on the other hand, limits this strategy to flat or slightly sloped tracks.

8

Open Problems and Conclusions

8.1 Conclusions

The increase of the engine output peak torque leads to the introduction of higher-capacity clutches. In the case of a manual transmission architecture this choice induces a higher force needed to operate the clutch pedal and a lower dosability. This trend, together with the reduction of the transmission stiffness in order to improve the NVH performances of the vehicle and the minimization of the friction losses to improve the fuel efficiency, can cause uncomfortable oscillations of the driveline that are both not sufficiently damped and difficult to avoid when performing a standing-start or a gearshift.

The solutions normally used to overcome these difficulties, namely the optimization of the clutch hydraulic control system to reduce the force necessary to operate the clutch pedal and the active damping of the driveline oscillations using the engine torque, do not show a large improvement margin and just simply keeping the actual comfort level on when designing a new vehicle can be challenging.

Since the manual transmission is still the default choice in Europe, the core market of Renault, the initial focus of this research has been gaining a better understanding of the comfort of a manual transmission clutch and eventually finding new means of improving it.

Following the traditional approach of optimizing the clutch hydraulic control circuit, initially we investigated the possibility of improving the clutch comfort through simple passive mechanical means. The results of this analysis show that any filtering action on the clutch pedal position hinders the driver and reduces the perceived comfort.

An optimization of the curve of the transmissible torque as a function of the clutch pedal position, obtained through a variable reduction-ratio system, has also been considered. This solution, in principle similar to the saturated flat spring that has already been studied at Renault, needs active assistance for reducing the clutch pedal activation force.

The clutch pedal activation force is an important aspect of the clutch comfort that has not been taken into consideration in this research work. Some active force reduction systems are proposed by tier-one suppliers but, due to their cost, have been very rarely used on commercial vehicles.

The introduction of an active element in the clutch control system opens the possibility of several innovative solutions for improving the clutch comfort through a careful control of this additional degree of freedom. The strategy presented in Chapter 3 is the result of the research in this direction and can be implemented either on a manual transmission equipped with a system partially decoupling the washer-spring's fingers position from the clutch pedal position or on a completely decoupled system like a clutch by wire or an automated manual transmission.

The experimental results obtained on a prototype vehicle equipped with an automated manual transmission have shown the actual comfort increase induced by this strategy. The introduction of a partially decoupling system on a manual transmission is, however, economically hardly justifiable at the moment.

The estimated mass-production cost of an active pedal activation force-reduction assistance is normally lower than the cost of a complete automated manual transmission module. Since a partially decoupling system would share much of the components of an automated manual transmission its final cost would probably fall in between the two previous systems. In the case of the active assistance and the partially decoupling system this additional cost does not induce a new function easily perceived by the driver/buyer of the vehicle. An automated manual transmission, instead, can be sold as an optional equipment and the induced increase in comfort relative to a normal manual transmission is easily perceived.

Beyond the technical difficulties specific to the manual transmission and economic reasons a more broader in the general public perception of the car object can accelerate the transition to partially or completely automated transmission. The car itself, in fact, is increasingly less associated with a mental image of exceptional moments, strong feelings and control of the vehicle and more thought of as an everyday object assuring a transport function; the main focus is less on the driving pleasure and more about the comfort. Inside this broad shift advanced control strategies, such as those presented in Chapters 3 and 4, could make the transition from the manual transmission easier in a more conservative market like the European one. These strategies help, together with the introduction of steering wheel paddles for triggering the gearshifts, change the perception of the automated manual transmission from an economic replacement of a traditional fully automatic transmission to a high-performance partially manual transmission. This perspective change is also, obviously, assisted by the adoption of this kind of transmission on Formula 1 and rally racing cars.

8.2 Further Work

The synchronization assistance strategy has been successfully tested for standing-starts on a flat ground. In order to make it more robust and allow standing-starts on steep slopes a feedback on the engine speed should be added. Moreover the extension to a gearshift should be fairly straightforward since the driver's input is limited to the gearshift triggering signal.

The simplified MPC control strategy has been only partially implemented but preliminary results seem to show that the computational level required by the strategy is compatible with an online implementation. This research direction is interesting because it allows the open-loop phase that needs extensive and time consuming tuning, but still requires some in depth analysis both of the theoretical and practical implementation aspects to be avoided.

Concerning the manual transmission, on the other hand, the main improving direction this research has highlighted is a better engine control based on an estimation of the transmitted torque by the clutch. This estimation could be used not only for improving the standing-start performance but also to increase the speed and comfort of the gearshifts by matching the engine torque to the transmitted clutch torque.

Appendix A

Optimization Methods

A.1 Dynamic Lagrangian Multipliers

The dynamic Lagrangian multipliers method is an extension of the Lagrangian multipliers method for dynamic optimization.

A dynamic optimization problem consists in finding the function $u(t)$ over the time interval T, eventually infinite, which minimizes the functional

$$J[u] = \int_T \mathcal{L}(x, u)\mathrm{d}t, \tag{A.1}$$

under the constraints of the differential equation defining the system dynamics and, optionally, prescribed initial and/or final states and additional inequality constraints.

The basic principle is to define using some additional variables, called Lagrangian multipliers, a new Lagrangian function $\mathcal{L}'$ that embeds the constraints. The solution of the new unconstrained optimization problem, called the *dual problem*, is, at worst, an upper bound of the solution of the original constrained problem and is actually coincident with that of the original problem if the so-called *strong duality* property is verified, as in the case of convex optimization problems.

The dual problem is solved by imposing the KKT (Karush Kuhn Tucker) optimality conditions. Since the constraints we are considering are linear, thanks to Abadie's constraint qualification, the KKT conditions are necessary and sufficient, *i.e.* every point satisfying the KKT conditions is a solution of the optimization problem and *vice versa*. The KKT conditions, on the other hand, are not always constructive meaning, that their equations do not always completely define the solution.

In the following we will detail the application of this method to a finite time dynamic optimization problem with prescribed initial and final states both with and without additional inequality conditions. In the first case the KKT conditions form a well-defined solution, while in the latter case the solution is not fully defined.

A.1.1 Inequality Constraints-free Optimization

The optimization problem under consideration is finding the function $u(t)$ over the time interval $T = [t_0, t_s]$ that minimizes the functional

$$J[u] = \int_{t_0}^{t_f} \mathcal{L}(x, u)dt,$$

where

$$\mathcal{L}(x, u) = \frac{1}{2} \left(x^T \mathbf{Q} x + u^T \mathbf{R} u \right) \tag{A.2}$$

under the following constraints

$$\dot{x} = f(x, u) = \mathbf{A}x + \mathbf{B}u \tag{A.3}$$

$$x(t_0) = x_0 \quad x(t_f) = x_f.$$

As anticipated, a new Lagrangian function is defined

$$\mathcal{L}'(x, \dot{x}, u, \lambda) = \mathcal{L}(x, u) + \lambda^T \left(f(x, u) - \dot{x} \right),$$

defining a new unconstrained dynamic optimization problem: find u and λ that minimize the functional

$$\hat{J}[u, \lambda] = \int_{t_0}^{t_f} \mathcal{L}'(x, \dot{x}, u, \lambda)dt$$

such that

$$x(t_0) = x_0 \quad x(t_f) = x_f.$$

Considering the functional variation between $\hat{J}[x(t), u(t)]$ and $\hat{J}[x(t) + h_x(t), u(t) + h_u(t)]$ we have

$$\Delta \hat{J} = \int_{t_0}^{t_f} \left[\mathcal{L}'(x + h_x, \dot{x} + \dot{h}_x, u + h_u, \lambda) + \mathcal{L}'(x, \dot{x}, u, \lambda) \right] dt,$$

using a truncated Taylor series we can write

$$\delta \hat{J} = \int_{t_0}^{t_f} \left[\frac{\partial \mathcal{L}'}{\partial x} h_x + \frac{\partial \mathcal{L}'}{\partial \dot{x}} \dot{h}_x + \frac{\partial \mathcal{L}'}{\partial u} h_u \right] dt.$$

Integration by parts of the second term gives

$$\delta \hat{J} = \int_{t_0}^{t_f} \left(\frac{\partial \mathcal{L}'}{\partial x} + \frac{d}{dt} \frac{\partial \mathcal{L}'}{\partial \dot{x}} \right) h_x dt + \int_{t_0}^{t_f} \frac{\partial \mathcal{L}'}{\partial u} h_u dt + \frac{\partial \mathcal{L}'}{\partial u} h_u \Big|_{t_0}^{t_f} \qquad (A.4)$$

where

$$\left(\frac{\partial \mathcal{L}'}{\partial x} + \frac{d}{dt} \frac{\partial \mathcal{L}'}{\partial \dot{x}} \right) h_x \Big|_{t_0}^{t_f} = 0$$

since $h_x(t_0) = h_x(t_s) = 0$ due to the final and initial states constraint. As $h_x(t)$ and $h_u(t)$ are arbitrary

$$\frac{\partial \mathcal{L}'}{\partial x} + \frac{d}{dt} \frac{\partial \mathcal{L}'}{\partial \dot{x}} = 0 \qquad (A.5)$$

$$\frac{\partial \mathcal{L}'}{\partial u} = 0.$$

By simple calculation

$$\frac{\partial \mathcal{L}'}{\partial x} = \frac{\partial \mathcal{L}'}{\partial x} + \lambda^T \frac{\partial f}{\partial x} = x^T \mathbf{Q} + \lambda^T \mathbf{A}$$

$$\frac{\partial \mathcal{L}'}{\partial \dot{x}} = -\lambda^T$$

$$\frac{\partial \mathcal{L}'}{\partial u} = \frac{\partial \mathcal{L}'}{\partial u} + \lambda^T \frac{\partial f}{\partial u} = u^T \mathbf{R} + \lambda^T \mathbf{B}$$

we have a differential equation defining the evolution of the Lagrangian multipliers and the relationship between the optimal solution and the Lagrangian multipliers.

$$\dot{\lambda} = -\mathbf{Q}x - \mathbf{A}^T \lambda \qquad (A.6)$$

$$u = -\mathbf{R}^{-1} \mathbf{B}^T \lambda. \qquad (A.7)$$

These two equations, called *secondary KKT conditions*, together with the original system dynamic Equation A.3, called *primary KKT condition*, define a two-point boundary-value problem (TPBVP).

A.1.2 Optimization Under Inequality Constraints

We consider the problem of finding the function $u(t)$ over the time interval $T = [t_0, t_s]$ that minimizes

$$J[u] = \int_{t_0}^{t_f} \mathcal{L}(x, u) dt,$$

where
$$\mathcal{L}(x, u) = \frac{1}{2} \left(x^T \mathbf{Q} x + u^T \mathbf{R} u \right)$$

under the constraints
$$\dot{x} = f(x, u) = \mathbf{A}x + \mathbf{B}u$$
$$x(t_0) = x_0 \quad x(t_f) = x_f$$
$$h(x, u) = \begin{bmatrix} u \\ -\mathbf{C}x \end{bmatrix} \leq 0.$$

As in the previous case we define a new Lagrangian function
$$\mathcal{L}'(x, \dot{x}, u) = \mathcal{L}(x, u) + \lambda^T (f(x, u) - \dot{x}) + \mu^T (h(x, u) + s^2)$$

defining the dual problem. The embedding of the inequality constraints introduces a new series of Lagrange multipliers μ and ancillary variables s called *slack variables* that transform the inequality constraints in equality constraints.

Equations A.1.1, A.1.1 and A.4 still hold true and give Equation A.5. Calculating the partials
$$\frac{\partial \mathcal{L}'}{\partial x} = \frac{\partial \mathcal{L}'}{\partial x} + \lambda^T \frac{\partial f}{\partial x} + \mu^T \frac{\partial h}{\partial x} = x^T \mathbf{Q} + \lambda^T \mathbf{A} + \mu^T \begin{bmatrix} 0 \\ -\mathbf{C} \end{bmatrix}$$

$$\frac{\partial \mathcal{L}'}{\partial \dot{x}} = -\lambda^T$$

$$\frac{\partial \mathcal{L}'}{\partial u} = \frac{\partial \mathcal{L}'}{\partial u} + \lambda^T \frac{\partial f}{\partial u} + \mu^T \frac{\partial h}{\partial u} = u^T \mathbf{R} + \lambda^T \mathbf{B} + \mu^T \begin{bmatrix} 1 \\ 0 \end{bmatrix}$$

we have the *KKT secondary conditions*
$$\dot{\lambda} = -\mathbf{Q}x - \mathbf{A}^T \lambda - \begin{bmatrix} 0 & -\mathbf{C} \end{bmatrix} \mu$$
$$u = -\mathbf{R}^{-1}\mathbf{B}^T \lambda - \mathbf{R}^{-1} \begin{bmatrix} 1 & 0 \end{bmatrix} \mu$$
$$\mu \geq 0$$

which, together with the so-called *complementary slackness* condition
$$\mu^T \begin{bmatrix} u \\ -\mathbf{C}x \end{bmatrix} = 0$$

and the *KKT primary conditions*
$$\dot{x} = f(x, u) = \mathbf{A}x + \mathbf{B}u$$
$$h(x, u) = \begin{bmatrix} u \\ -\mathbf{C}x \end{bmatrix} \leq 0,$$

are the set of KKT conditions for the dual problem. The additional Lagrange multipliers μ are not completely defined by the previous equations. In this case, the KKT conditions are not constructive and do not define a solution to the optimization problem that has been solved using a quadratic programming formulation.

A.2 Alternative Solution of the Two-point Boundary-value Problem by Generating Functions

The optimal control $u(t)$ for the inequality constraint-free optimization problem is defined by the solution of the following TPBVP

$$\dot{x} = \mathbf{A}x + \mathbf{B}_e \Gamma_e + \mathbf{B}_c u \qquad (A.8a)$$

$$\dot{\lambda} = -\mathbf{Q}x - \mathbf{A}^T \lambda \qquad (A.8b)$$

$$u = -\mathbf{R}^{-1}\mathbf{B}_c^T \lambda \qquad (A.8c)$$

given by the *primary KKT condition* (A.3) and the two *secondary KKT conditions*.

Two solutions to this problem have been presented in Chapter 3: the iterative shooting method and the analytical solution using a matrix exponential. An alternative solution of the TPBVP uses the properties of the Hamiltonian systems, namely the canonic transformations defined by a generating function, to obtain the initial co-state vector λ_0. This approach, quite complex on both the theoretical and practical planes, is the only available solution to finite-time optimal control problems over very long time intervals. The most frequent example of these systems found in the literature is the optimal orbit change and *rendezvous* planning for satellites having a propulsion system too weak to use the impulsive speed change approximation used in Boltzmann orbits.

The previously defined TPBVP can be written as an Hamiltonian system in homogeneous canonic form plus the forced reaction to the engine torque Γ_e

$$\dot{z} = \begin{bmatrix} \dot{x} \\ \dot{\lambda} \end{bmatrix} = \begin{bmatrix} \frac{\partial H(y,\lambda,u)}{\partial \lambda} \\ \frac{\partial H(y,\lambda,u)}{\partial y} \end{bmatrix} + \mathbf{B}_e \Gamma_e,$$

where $H(x, \lambda, u) = \mathcal{L}(x, u) + \lambda^T(\mathbf{A}x + \mathbf{B}_c u)$ is the Hamiltonian; $\mathcal{L}(x, u)$ is the Lagrangian defined by Equation A.2. $x \in \mathbb{R}^n$ is the state vector of the system subject to the optimal control and $\lambda \in \mathbb{R}^n$ the corresponding co-states or dynamic Lagrangian multipliers.

The TPBVP resolution method proposed in [28] and [18] is defined only for homogeneous Hamiltonian systems, *i.e.* for control problems where all the system inputs are controlled inputs.

By linearity, we can separate the free evolution of the system $z_{free}(t)$ from the forced one $z_{\Gamma_e}(t)$

$$z(t) = e^{\mathbf{A}t}z_0 + \int_{t_0}^{t} e^{\mathbf{A}(\tau-t)}\mathbf{B}_e \Gamma_e(\tau)\mathrm{d}\tau = z_{free}(t) + z_{\Gamma_e}(t).$$

Since the engine torque Γ_e is assumed to be independent of the system evolution the forced evolution of the system can be simply calculated by forward integration before the solution of the optimal control problem.

The initial co-state vector $\lambda(t_0) = \lambda_0$, solution of the TPBVP

$$\dot{z} = \mathbf{A}z + \mathbf{B}_e \Gamma_e$$
$$x(t_0) = x_0 \quad x(t_s) = x_s$$

is also the solution of the homogeneous TPBVP

$$\dot{z} = \mathbf{A}z$$
$$x(t_0) = x_0 \quad x(t_s) = x_s - x_{\Gamma_e}(t_s),$$

where

$$z_{\Gamma_e}(t_s) = \int_{t_0}^{t_s} e^{\mathbf{A}(\tau - t_s)} \mathbf{B}_e \Gamma_e(\tau) d\tau = \left[x_{\Gamma_e}(t_s) \ \lambda_{\Gamma_e}(t_s) \right]^T.$$

For a better understanding of the TPBVP solution method by generating functions the Hamilton principle, also called the minimum effort principle, and the definition of a canonic transformation between extended phase spaces are briefly recalled.

Definition A.1 (Hamilton Principle). *The trajectory of a Hamiltonian system in the phase space makes the following integral extremal*

$$\int_{t_0}^{t_f} [\lambda \dot{y}^T - H(y, \lambda, t)] dt = \int_{t_0}^{t_f} \mathcal{L}(x, u) dt,$$

which implies

$$\delta \int_{t_0}^{t_f} \mathcal{L}(x, u) dt = 0.$$

Definition A.2 (Extended Phase Space). *Let $P \in \mathbb{R}^{2n}$ be a phase space, $P \times \mathbb{R}$ is called an extended phase space (by time).*

Definition A.3 (Canonic Transformation). *A map $f : P_1 \times \mathbb{R} \to P_2 \times \mathbb{R}$ is said to be a canonic transformation if:*

- *f is an isomorphism C^∞;*
- *f does not affect time, i.e. $\exists g_T(z)$ such that $f(z, t) = (g_T(z), t)$; and*
- *f preserves the canonic form of the Hamiltonian systems.*

The last point is equivalent to assuring that there exists a Hamiltonian K after the transformation $(X, \Lambda) = f(x, \lambda)$ such that the system dynamics can be written as

$$\begin{bmatrix} \dot{X} \\ \dot{\Lambda} \end{bmatrix} = \begin{bmatrix} \frac{\partial K}{\partial \Lambda} \\ \frac{\partial K}{\partial Y} \end{bmatrix}.$$

A.2.1 Generating Functions

The Hamilton principle, invariant for canonical transformations, implies that

$$\lambda \dot{x} - H = \Lambda \dot{X} - K + \frac{\mathrm{d}F}{\mathrm{d}t}, \tag{A.9}$$

where F is a function, called the generating function, defining the canonical transformation. In principle, this function depends on $4n+1$ parameters, *i.e.* x, λ, X, Λ and t but because of the $2n$ constraints imposed by the canonical transformation f, F is a function of just $2n+1$ parameters. Amongst all the possible choices for the set of independent parameters there are four classic formulations

$$F_1(x, X, t) \quad F_2(x, \Lambda, t) \quad F_3(\lambda, X, t) \quad F_4(\lambda, \Lambda, t).$$

Calculating the total derivative $\mathrm{d}F/\mathrm{d}t$ for the first two classical formulations and substituting the result in Equation A.9 we have, under the hypothesis of independent parameters

$$\lambda = \frac{\partial F_1(x,X,t)}{\partial x} \qquad\qquad \lambda = \frac{\partial F_2(x,\Lambda,t)}{\partial x}$$

$$\Lambda = -\frac{\partial F_1(x,X,t)}{\partial X} \qquad\qquad X = -\frac{\partial F_2(x,\Lambda,t)}{\partial \Lambda} \qquad .$$

$$H(x,\lambda,t) + \frac{\partial F_1(x,X,t)}{\partial t} = K(X,\Lambda,t) \quad H(x,\lambda,t) + \frac{\partial F_2(x,\Lambda,t)}{\partial t} = K(X,\Lambda,t)$$

The equations in the last line are known as Hamilton-Jacobi PDEs; their solution allows the generating function to be obtained. Once the generating function has been obtained in one of the four classical formulations the other three can be calculated by applying a Legendre transformation.

A.2.2 Hamiltonian System Flow

The flow $\phi : (y(t_0), \lambda(t_0), t) \rightarrow (y(t), \lambda(t), t)$ of the Hamiltonian system is a canonical transformation. For a linear system this flow is usually expressed using a matrix exponential

$$\begin{bmatrix} x(t) \\ \lambda(t) \end{bmatrix} = \mathrm{e}^{A_L(t-t0)} \begin{bmatrix} x_0 \\ \lambda_0 \end{bmatrix} .$$

The interest of the TPBVP solution by means of the generating functions is to obtain the same relationship without using the matrix exponential that causes some numerical difficulties. This method can also be used to obtain an approximated analytic solution of finite-time optimal control for non-linear systems.

Using the formalism introduced in the definition of a canonical transformation we have

$$X = x(t_0) = x_0 \quad \Lambda = \lambda(t_0) = \lambda_0$$

giving the corresponding Hamiltonian system

$$\begin{bmatrix} \dot{x}_0 = 0 \\ \dot{\lambda}_0 = 0 \end{bmatrix} = \begin{bmatrix} \frac{\partial K(x_0, \lambda_0, t)}{\partial \lambda_0} \\ \frac{\partial K(x_0, \lambda_0, t)}{\partial x_0} \end{bmatrix}.$$

Since K is constant we can assume $K \equiv 0$ without any loss of generality.

A.2.3 Two-point Boundary-value Problem Solution

The generating function first classical formulation $F_1(x, x_0, t)$ is the most apt to solve the TPBVP since

$$\lambda_0 = \left. \frac{\partial F_1(x, x_0, t)}{\partial x} \right|_{x = x_f \; t = t_f}. \tag{A.10}$$

Unluckily this generating function cannot be directly obtained from the Hamilton Jacobi equation since in t_0 the assumption of independence of the parameters is not verified: $F_1(x_0, x_0, t_0)$.

The solution to this difficulty proposed in [18] consists in obtaining first the generating function in its second classical formulation $F_2(x, \lambda_0, t)$, which satisfies the independence assumption in t_0, from the Hamilton Jacobi equation

$$H(x, \lambda, t) + \frac{\partial F_2(x, \lambda_0, t)}{\partial t} = 0,$$

and then obtain $F_1(x, x_0, t_0)$ through a Legendre transformation

$$F_1(x, x_0, t) = F_2(x, \lambda_0, t) - x_0^T \lambda_0, \tag{A.11}$$

and finally obtain the initial co-states vector λ_0.

The finite time optimal control of a linear system

$$\dot{x} = \mathbf{A}x + \mathbf{B}u$$

with respect to the quadratic cost function

$$J[u] = \int_{t_0}^{t_s} \left[x^T \mathbf{Q} x + u^T \mathbf{R} u \right] dt$$

induces a quadratic Hamiltonian

$$H(x, \lambda, t) = \frac{1}{2} \begin{bmatrix} x \\ \lambda \end{bmatrix}^T \begin{bmatrix} \mathbf{Q} & \mathbf{A}^T \\ \mathbf{A} & -\mathbf{B}\mathbf{R}^{-1}\mathbf{B}^T \end{bmatrix} \begin{bmatrix} x \\ \lambda \end{bmatrix},$$

which allows F_2 to be written as

$$F_2(x, \lambda_0, t) = \frac{1}{2} \begin{bmatrix} x \\ \lambda_0 \end{bmatrix}^T \begin{bmatrix} \mathbf{F}_{xx} & \mathbf{F}_{x\lambda_0} \\ \mathbf{F}_{x\lambda_0}^T & \mathbf{F}_{\lambda_0\lambda_0} \end{bmatrix} \begin{bmatrix} x \\ \lambda_0 \end{bmatrix}. \tag{A.12}$$

Substituting Equation A.12 in the PDE Hamilton Jacobi equation relative to F_2 we obtain a system of matrix differential equations

$$\dot{\mathbf{F}}_{xx} + \mathbf{Q} + \mathbf{F}_{xx}\mathbf{A} + \mathbf{A}^T\mathbf{F}_{xx} - \mathbf{F}_{xx}\mathbf{B}\mathbf{R}^{-1}\mathbf{B}^T\mathbf{F}_{xx} = 0 \tag{A.13a}$$

$$\dot{\mathbf{F}}_{x\lambda_0} + \mathbf{A}^T\mathbf{F}_{x\lambda_0} - \mathbf{F}_{xx}\mathbf{B}\mathbf{R}^{-1}\mathbf{B}^T\mathbf{F}_{x\lambda_0} = 0 \tag{A.13b}$$

$$\dot{\mathbf{F}}_{\lambda_0\lambda_0} - \mathbf{F}_{x\lambda_0}\mathbf{B}\mathbf{R}^{-1}\mathbf{B}^T\mathbf{F}_{x\lambda_0} = 0, \tag{A.13c}$$

having as initial conditions

$$\mathbf{F}_{xx}(t_0) = 0_{n \times n} \quad \mathbf{F}_{x\lambda_0}(t_0) = I_{n \times n} \quad \mathbf{F}_{\lambda_0\lambda_0}(t_0) = 0_{n \times n}.$$

Once the matrix differential system (A.13) is resolved the F_1 formulation of the generating function is obtained by the Legendre transformation (A.11). Finally, thanks to the relation (A.10), we have

$$\lambda_0 = \mathbf{F}_{\lambda_0\lambda_0}^{-1}(t_s) \left[x_0 - \mathbf{F}_{\lambda_0 x}(t_s)x_s \right].$$

This solution is numerically stable but quite complex since it requires the integration of a system of 48 differential equations. In the case of a clutch optimal engagement control the solution by quadratic programming formulation is still feasible thanks to a relatively short optimization horizon. This solution, which is both simpler and more powerful since it allows the inclusion of additional inequality constraints, has been chosen as the standard solution of the optimization problem.

A.3 Reconduction to a Quadratic Programming Formulation

The optimization program posed by the optimal engagement control is to find the function $u(t)$ over the interval $T = [t_0, t_s]$ minimizing the functional

$$J[u] = \int_{t_0}^{t_f} \mathcal{L}(x, u)\mathrm{d}t,$$

where

$$\mathcal{L}(x, u) = \frac{1}{2} \left(x^T\mathbf{Q}x + u^T\mathbf{R}u \right)$$

under the constraints

$$\dot{x} = f(x, u) = \mathbf{A}x + \mathbf{B}u \tag{A.14}$$

$$x(t_0) = x_0 \quad x(t_f) = x_f \tag{A.15}$$
$$\mathbf{A}_c x \le \mathbf{B}_c.$$

The sampling of the dynamic of the system subject to the optimal control is used to reduce the dynamic optimization to a quadratic program, *i.e.* the optimization of a vector composed by the samples u_k of the function $u(t)$ taken at the sampling instants t_k. The solution to the optimization problem is, thus, the vector

$$\bar{u} = \begin{bmatrix} u_0 & u_1 & \dots & u_{N-1} \end{bmatrix}^T,$$

where N is the number of samples over the optimization horizon T.

Iterating the finite-difference equation of the sampled system

$$x_{k+1} = A_d x_k + B_d u_k,$$

we have the following relation

$$x_k = A_d^k x_0 + A_d^{k-1} B_d u_0 + A_d^{k-2} B_d u_1 + \dots + A_d B_d u_{k-2} + B_d u_{k-1}$$

defining the sample x_k as a function of the initial state x_0 and the input samples u_i with $i \in [0, k-1]$.

This relation can be put in matrix form

$$\begin{bmatrix} x_1 \\ x_2 \\ \vdots \\ x_N \end{bmatrix} = \begin{bmatrix} B_d & 0 & \cdots & 0 \\ A_d B_d & B_d & \cdots & 0 \\ \vdots & \vdots & \ddots & \vdots \\ A_d^{N-1} B_d & A_d^{N-2} B_d & \cdots & B_d \end{bmatrix} \begin{bmatrix} u_0 \\ u_1 \\ \vdots \\ u_{N-1} \end{bmatrix} + \begin{bmatrix} A_d \\ A_d^2 \\ \vdots \\ A_d^N \end{bmatrix} x_0$$

which, in a more compact form, is written as

$$\bar{x} = E\bar{u} + Fx_0. \tag{A.16}$$

The previous equation expresses the vector $\bar{x}$ formed by the sampled state vectors as a function of the initial state vector x_0 and the vector $\bar{u}$.

Due to the sampling the integral functional is simplified in a sum

$$J[\bar{u}] = \sum_{k=1}^{N} x_k^T Q x_k + \sum_{k=0}^{N-1} u_k^T R u_k,$$

which can be expressed using the vectors $\bar{x}$ and $\bar{u}$

$$J[\bar{u}] = \begin{bmatrix} x_1 & \dots & x_N \end{bmatrix} \begin{bmatrix} \mathbf{Q} & \cdots & 0 \\ \vdots & \ddots & \vdots \\ 0 & \cdots & \mathbf{Q} \end{bmatrix} \begin{bmatrix} x_1 \\ \cdots \\ x_N \end{bmatrix}$$
$$+ \begin{bmatrix} u_0 & \dots & u_{N-1} \end{bmatrix} \begin{bmatrix} \mathbf{R} & \cdots & 0 \\ \vdots & \ddots & \vdots \\ 0 & \cdots & \mathbf{R} \end{bmatrix} \begin{bmatrix} u_0 \\ \cdots \\ u_{N-1} \end{bmatrix} \tag{A.17}$$
$$= \bar{x}^T \bar{\mathbf{Q}} \bar{x} + \bar{u}^T \bar{\mathbf{R}} \bar{u}.$$

Substituting Equation A.16 in Equation A.17 we have, finally, a cost function in the standard QP formulation

$$J = \bar{u}^T \left(E^T \bar{Q} E + \bar{R} \right) \bar{u} + x_0 F^T \bar{Q} E \bar{u}.$$

The equality constraint Equation A.14 due to the system dynamic has been embedded, through substitution, in the cost function. We still have to include the initial and final states constraints together with the inequality constraints.

From the last line of the Equation A.16 we have

$$x_s = \left[\mathbf{A}_d^{N-1} \mathbf{B}_d \; \cdots \; \mathbf{B}_d \right] \bar{u} + \mathbf{A}^N x_0,$$

which in standard representation gives

$$\left[\mathbf{A}_d^{N-1} \mathbf{B}_d \; \cdots \; \mathbf{B}_d \right] \bar{u} = x_s - \mathbf{A}^N x_0$$

$$\mathbf{A}_{eq} \bar{u} = b_{eq}.$$

The inequality constraints in matrix form become

$$\begin{bmatrix} \mathbf{A}_c & \cdots & 0 \\ \vdots & \ddots & \vdots \\ 0 & \cdots & \mathbf{A}_c \end{bmatrix} \begin{bmatrix} x_1 \\ \cdots \\ x_N \end{bmatrix} \leq \begin{bmatrix} b_c \\ \cdots \\ b_c \end{bmatrix}$$

$$\bar{\mathbf{A}}_c \bar{x} \leq \bar{b}_c$$

substituting Equation A.16 we finally have

$$\bar{\mathbf{A}}_c E \bar{u} \leq \bar{b}_c$$

$$\mathbf{A}_{in} \bar{u} \leq b_{in}.$$

The sampling thus reduces the dynamic optimization into a static optimization that can be written in the standard QP formulation:

Find the vector $\bar{u}$ minimizing

$$J = \bar{u}^T \left(E^T \bar{Q} E + \bar{R} \right) \bar{u} + x_0 F^T \bar{Q} E \bar{u},$$

under the constraints

$$\boldsymbol{A}_{eq} \bar{u} = b_{eq}$$

$$\boldsymbol{A}_{in} \bar{u} \leq b_{in}.$$

Appendix B

Proof of Theorem 6.1

Lemma B.1. *The linear time-invariant system:*

$$\dot{x} = \boldsymbol{A}x + \boldsymbol{B}u \tag{B.1a}$$

$$y = \boldsymbol{C}x + \boldsymbol{D}u \tag{B.1b}$$

is finite gain $\mathcal{L}_p$ stable for every $p \in [1, \infty]$ if $\boldsymbol{A}$ is Hurwitz. Furthermore, the inequality relation

$$\|y\|_{\mathcal{L}_p} \leq \gamma \|u\|_{\mathcal{L}_p} + \beta$$

is verified for:

$$\gamma = \|\boldsymbol{D}\|_2 + \frac{2\lambda_{max}^2(\boldsymbol{P})\|\boldsymbol{B}\|_2\|\boldsymbol{C}\|_2}{\lambda_{min}(\boldsymbol{P})} \tag{B.2}$$

$$\beta = \rho\|\boldsymbol{C}\|_2\|x_0\|\sqrt{\frac{\lambda_{max}(\boldsymbol{P})}{\lambda_{min}(\boldsymbol{P})}} \tag{B.3}$$

$$\rho = \left\{ \begin{array}{ll} 1, & \text{if } p = \infty \\ \left(\frac{2\lambda_{max}(\boldsymbol{P})}{p}\right)^{1/p}, & \text{if } p \in [1, \infty) \end{array} \right\}$$

where $\boldsymbol{P}$ is the solution of the Riccati equation $\boldsymbol{P}\boldsymbol{A} + \boldsymbol{A}^T\boldsymbol{P} = -\boldsymbol{I}$.

Proof. This Lemma is the Corollary 5.2 of Theorem 5.1, the interested reader can find its proof on page 202 of [20].

Proof. By simple substitution

$$\dot{\tilde{x}} = (\mathbf{A} - \mathbf{KC})\tilde{x} + \mathbf{W}_1\epsilon_1 - \mathbf{KW}_2\epsilon_2.$$

By hypothesis, $\mathbf{A} - \mathbf{KC}$ is diagonalizable, *i.e.* a base change exists

$$\tilde{x} = \mathbf{T}z \tag{B.4}$$

such that

$$\dot{z} = \mathbf{D}z + \bar{\mathbf{B}}_1\epsilon_1 + \bar{\mathbf{B}}_2\epsilon_2, \tag{B.5}$$

where $\mathbf{D} = \mathbf{T}^{-1}(\mathbf{A} - \mathbf{KC})\mathbf{T}$ is a diagonal matrix, $\bar{\mathbf{B}}_1 = \mathbf{T}^{-1}\mathbf{W}_1$ and $\bar{\mathbf{B}}_2 = \mathbf{T}^{-1}\mathbf{KW}_2$. Furthermore, $\mathbf{T}$ has as its columns the eigenvectors of $A - KC$ that are defined but for a multiplicative constant. This degree of liberty allows $\|\mathbf{T}\|_2 = 1$ to be assumed without any loss of generality. Since $\mathbf{P} = -2\mathbf{D}^{-1}$ is the solution of the Lyapunov equation $\mathbf{D}^T\mathbf{P} + \mathbf{PD} = -\mathbf{I}$ we have the following relation

$$\lambda_{max}(P) = -1/(2\lambda_{min}(A - KC)) \tag{B.6a}$$
$$\lambda_{min}(P) = -1/(2\lambda_{max}(A - KC)). \tag{B.6b}$$

Using the superposition principle, the Lemma B.1 and the relation between the eigenvalues (B.6) to the system (B.5) we have:

$$\|z\|_{\mathcal{L}_p} \leq \gamma_1\|\epsilon_1\|_{\mathcal{L}_p} + \gamma_2\|\epsilon_2\|_{\mathcal{L}_p} + \beta \tag{B.7}$$

$$\gamma_1 = -\frac{\lambda_{max}}{\lambda_{min}^2}\|\bar{B}_1\|_{\mathcal{L}_p} \quad \gamma_2 = -\frac{\lambda_{max}}{\lambda_{min}^2}\|\bar{B}_2\|_{\mathcal{L}_p} \tag{B.8}$$

$$\beta = \rho\|z(0)\|\sqrt{\frac{\lambda_{max}}{\lambda_{min}}} \tag{B.9}$$

$$\lambda_{max} = \max\{\lambda(D)\} \tag{B.10}$$
$$\lambda_{min} = \min\{\lambda(D)\}. \tag{B.11}$$

Since $\lambda(D) = \lambda(A - KC)$ and $\|\tilde{x}\|_{\mathcal{L}_p} = \|Tz\|_{\mathcal{L}_p} \leq \|T\|_2\|z\|_{\mathcal{L}_p} = \|z\|_{\mathcal{L}_p}$ we have the thesis.

$\square$

Appendix C

Brief Description of the LuGre Model

The LuGre model is a dynamic friction model presented in [9]. Friction is modeled as the average deflection force of elastic springs. When a tangential force is applied the bristles will deflect like springs. If the deflection is sufficiently large the bristles start to slip. The average bristle deflection for a steady state motion is determined by the velocity. It is lower at low velocities, which implies that the steady state deflection decreases with increasing velocity. This models the phenomenon that the surfaces are pushed apart by the lubricant, and models the Stribeck effect. The model also includes rate dependent friction phenomena such as varying break-away force and frictional lag. The model has the form

$$\frac{\mathrm{d}z}{\mathrm{d}t} = v - \sigma_0 \frac{|v|}{g(v)} z$$

$$F = \sigma_0 z + \sigma_1(v) \frac{\mathrm{d}z}{\mathrm{d}t} + F(v)$$

where z denotes the average bristle deflection. The model behaves like a spring for small displacements. The parameter σ_0 is the stiffness of the bristles, and $\sigma_1(v)$ the damping. The function $g(v)$ models the Stribeck effect, and $f(v)$ is the viscous friction. A reasonable choice of $g(v)$ which gives a good approximation of the Stribeck effect is

$$g(v) = \alpha_0 + \alpha_1 e^{-(v-v_0)^2}$$

The sum $\alpha_0 + \alpha_1$ then corresponds to stiction force and α_0 to Coulomb friction force. The parameter v_0 determines how $g(v)$ varies within its bounds $\alpha_0 \leq g(v) \leq \alpha_0 + \alpha_1$. A common choice of $f(v)$ is linear viscous friction $f(v) = \alpha_2 v$.

For a more advanced analysis of this model please see [9] and [27].

References

1. S. K. Agrawal and B. C. Fabien. *Optimisation of Dynamic Systems*. Lecture Notes, 1994.
2. A. E. Anderson and R. A. Knapp. Hot spotting in automotive friction systems. *Proc. of International Conference on Wear of Materials*, 135:319–337, 1990.
3. J. R. Barber. Thermoelastic instabilities in the sliding of conforming solids. *Proc. R. Soc. Lond.*, A312:381–394, 1969.
4. G. Bastin and M. Gevers. Stable adaptive observers for nonlinear time varying systems. *IEEE Trans. Auto. Cont.*, 33(7):650–658, 1988.
5. A. Bemporad, F. Borrelli, L. Glielmo, and F. Vasca. Hybrid control of dry clutch engagement. *Proc. of European Control Conf. Porto*, 2001.
6. G. Besançon. Remarks on nonlinear adaptive observer design. *Systems Cont. Lett.*, 41(4):271–280, 2000.
7. P. A. Bliman, T. Bonald, and M. Sorine. Hysteresis operators and tire friction models: application to tire dynamic simulator. *Proc. of ICIAM Hamburg*, 1995.
8. C. Commault, J. M. Dion, O. Sename, and R. Monteyian. Unknown input observer - a structural approach. *Proc. ECC congress*, 2001.
9. C. Canudas de Wit, H. Olsson, K. J. Åström, and P. Lischinsky. A new model for control of systems with friction. *IEEE Trans. Auto. Cont.*, 40(3):419–425, March 1995.
10. P. J. Dolcini. Etude de dosabilité d'un embrayage. Technical report, Renault SAS, 2002.
11. P. J. Dolcini, C. Canudas de Wit, and H. Béchart. Improved optimal control of dry clutch engagement. *Proc. 16th IFAC World Conference Prague*, 2005.
12. P. J. Dolcini, C. Canudas de Wit, and H. Béchart. Observer based optimal control of dry clutch engagement. *Proc. CDC-ECC Joint Conference Seville*, 2005.
13. J. Fredriksson and B. Egardt. Nonlinear control applied to gearshifting in automated manual transmissions. *Proc. of the 39th IEEE CDC*, 1:444–449, 2000.
14. F. Garofalo, L. Glielmo, L. Iannelli, and F. Vasca. Smooth engagement of automotive dry clutch. *Proc. 40th IEEE CDC*, pages 529–534, 2001.
15. F. Garofalo, L. Glielmo, L. Iannellli, and F. Vasca. Optimal tracking for automotive dry clutch engagement. *Proc. 15th IFAC congress*, 2002.
16. L. Glielmo and F. Vasca. Optimal control of dry clutch engagement. *SAE*, (2000-01-0837), 2000.

17. G. C. Goodwind, M. S. Seron, and J. A. De-Dona. *Contrained Control and Estimation*. Springer Verlag (Germany), 2005.

18. V. M. Guibout and D. J. Scheeres. Formation flight with generating functions: Solving the relative boundary value problem. *AIAA Astrodynamics Specialist Meeting*, (AIAA 2002-4639), 2002.

19. P. O. Gutman and L. Glielmo. A literature study of clutch and driveline modeling and control. *Informal Workshop on Clutch Control, Benevento Italy*, 2006.

20. H. K. Khalil. *Nonlinear Systems Third Edition*. Patience Hall, 2002.

21. U. Kiencke and L. Nielsen. *Automotive Control Systems, For Engine, Driveline, and Vehicle 2nd Edition*. Springer Verlag (Germany), 2005.

22. G. Kreisselmeier. Adaptive observer with exponential rate of convergence. *IEEE Trans. Auto. Cont.*, 22(1):2–8, 1977.

23. R. Lindas. Embrayages - étude technologique. *Documentation technique Valeo B-5-851*.

24. R. Marino and P. Tomei. Adaptive obeservers with arbitrary exponential rate of convergence for nonlinear systems. *IEEE Trans. Auto. Cont.*, 40(7):1300–1304, 1995.

25. C. Moler and C. V. Loan. Nineteen doubious ways to compute the exponential of a matrix. *SIAM Rev.*, 20(4):801–836, 1978.

26. C. Moler and C. V. Loan. Nineteen doubious ways to compute the exponential of a matrix, twenty five years later. *SIAM Rev.*, 45(1):3–49, 2003.

27. H. Olsson, K. J. Åström, C. Canudas de Wit, M. Gäfvert, and P. Lischinsky. Friction models and friction compensation. *Euro. J. Control*, 4:176–195, 1997.

28. C. Park and D. J. Scheeres. Solutions of optimal feedback control problem with general boundary conditions using hamiltonian dynamics and generating functions. *Automatica*, 42:869–875, 2006.

29. M. Pettersson and L. Nielsen. Gear shifting by engine control. *IEEE Trans. Cont. Systems Technol*, Volume 8(No. 3):pp. 495–507, May 2000.

30. M. Pettersson and L. Nielsen. Diesel engine speed control with handling of driveline resonances. *Cont Eng Prac*, 11(No. 10):pp. 319–328, July 2003.

31. S. Shuiwen, G. Anlin, L. Bangjie, Z. Tianyi, and F. Juexin. The fuzzy control of a clutch of an electronically controlled automatic mechanical transmission. *JSAE Technical Paper Series*, (9530805), 1995.

32. J. Sliker and R. N. K. Loh. Design of robust vehicle launch control systems. *IEEE Trans. Cont. System Technol.*, 4(4):326–335, 1996.

33. H. Tanaka and H. Wada. Fuzzy control of engagement for automated manual transmission. *Vehicle System Dyn.*, 24:365–366, 1995.

34. C. Canudas De Wit, P. Tsiotras, E. Velenis, M. Basset, and G. Gissinger. Dynamic friction models for road/tire longitudinal interaction. *Vehicle System Dyn.*, 39(3):189–226, 2002.

35. Y. B. Yi, J. R. Barber, and P. Zagrodzki. Eigenvalue solution of thermoelastic instability problems using fourier reduction. *Proc. Roy. Soc. London*, A456:2799–2821, 2000.

36. Y. B. Yi, S. Du, J. R. Barber, and J. W. Fash. Effect of geometry on thermoelastic instability in disk brakes and clutches. *ASME J. of Tribo.*, 121:661–666, 1999.

37. Q. Zhang. Adaptative observer for mimo linear time varying systems. *INRIA research report - theme 4*, (INRIA/RR-4111-FR+ENG), 2001.

38. Q. Zhang and B. Delyon. A new approach to adaptive observer design for mimo systems. *Proc. ACC congress*, 2001.

Index

Other titles published in this series (continued):